Arctic

Forces

Tyler Blackthorne

A Dylan Baker Thriller

Mt. Hood Press
Portland, Oregon

Arctic Forces
A Mt. Hood Press Book
ISBN 978-0-9772608-5-0

Published by Mt. Hood Press
Portland, Oregon

Manufactured in the United States of America

Special thanks to readers, friends and family who offered valuable feedback and encouragement.

Sharon Hansen, Louise Young, Rick Young, Kelly Hansen, Mike Doran, Katy Henke, Brent Lossing, Debbie Robillard, Mary Brandl, Anita Bendickson, Richard Mace, Susan Mace, Susan Edmonson

Author's Note

Global warming concerns and the research efforts to repurpose carbon emissions are prevalent in both the daily news and technical literature. Enhanced by genetic engineering, microorganisms like *Ralstonia* and others are currently available to assist in this endeavor.

Although the weapons and the mechanical devices mentioned in ***Arctic Forces*** exist today, there are two devices (e.g., the Layer Cake Membrane and Sully's portable testing device) that have not yet reached the market place. For more complete information please visit my publisher's website, www.mthoodpress.com

Tyler Blackthorne

PROLOGUE

PROFESSOR NEIL CARLYLE examined the settings on his tripod-mounted-camera using the reflected glow from his red-light headlamp. He checked for the constellation Aquarius rising over Mt. Hood. This was the perfect night for star photography, no moon, Eta Aquarids meteor shower at full blast, and crystal clear skies at the snowy six thousand foot level.

He opened the shutter for a 20-second exposure, and then marveled at silver-gold meteor streaks over the cold beauty of Mt. Hood. Perfect!

Somewhere behind him, he heard a twig snap. Part of Professor Carlyle's mind wondered what would cause a stick to snap on a black, cold night illuminated by only star shine on snow. He had hiked for hours to get away from any light pollution and set up his winter campsite just at dusk.

As his camera shutter clicked, he heard a snow-laden branch give up its load with a deep thump. If someone else was up here, the professor hoped they'd have enough sense not to turn on any lights while his camera's shutter was open.

Dr. Carlyle moved his camera to capture the southern night sky. It seemed more of the meteor trails were appearing in that direction.

From out of the blackness, an authoritative female voice broke the silence, "Dr. Carlyle?"

Starting, Dr. Carlyle bumped his tripod sending the camera lens-down into a pile of rocks. "Damn! I nearly peed my pants. Look at my camera!" Carlyle reached to his forehead and switched on his dim red light. In the eerie glow, he could see two soldiers in Arctic combat gear standing right next to him. They were regarding him like robots through night-vision goggles.

"What's going on?" Dr. Carlyle asked as he picked up his tripod and examined the camera's lens. With only the faint red glow from his headlamp, he couldn't tell if the lens was scratched or just dirty.

"Dr. Carlyle? Dr. Neil Carlyle?" the female soldier's voice indicated she meant business.

"Yes. What's going on? Why are you here in the night asking for me? Is it Jennie?" Dr. Carlyle thought of his ailing wife at home with family in Eugene.

"Sir, you need to come with us." The female soldier took his arm and started to lead him out of the camp.

Pulling his arm out of her grip, Dr. Carlyle turned his headlamp to full white light flooding the area with brilliant LED-powered illumination. The two soldiers flinched in the light. Carlyle could see four other well-armed soldiers back in the woods.

"What's going on? Why are you here?" Carlyle was obviously frightened and confused.

Snapping their NV goggles up, the soldiers raised their rifles towards him. "Dr. Carlyle, you are interfering with a military exercise. You need to come with us." This time the soldier's painful grip on his arm made it clear, he wouldn't be able to just pull free.

"Military exercise? On National Recreational Land? I don't think so. You guys are not allowed to train up here." From behind, a large male soldier with long blond-red hair cascading from under his helmet, cuffed Dr. Carlyle's hands behind his back.

"What the . . .?" Carlyle struggled.

The female soldier shoved a large roll of gauze into his mouth and taped his face so he couldn't talk.

They led the bewildered professor, not back to the parking lot, but up the mountain to a steep hillside. Here the cuffs and gag came off. Carlyle looked around to see soldiers near the hillside were arranging his tent and other camping gear.

"Reno, get him into the tent," the female solder ordered.

Anyone observing could tell the professor was on the verge of panic. Matter-of-factly, Reno slammed a chunk of ice into the side of Carlyle's head. Groaning, Carlyle went down to his knees. Two soldiers shoved the semi-conscious man into his tent, and then ran back the way they came.

At a signal, a powerful blast exploded up the mountain, followed by the roar of heavy snow and rock sliding downwards. Moments later, the tent containing Carlyle, vanished under 30 feet of avalanche debris.

"Tango, what about our footprints?" Reno asked.

"He told everyone not to expect him for two or so days. Snow showers are forecast for the morning. We're done here."

DYLAN COULDN'T REMEMBER EVER being so uneasy as he gazed out the windows of his hand-built log home. Even deep in the woods, the surrounding Kenai Fjords of Alaska couldn't seem to calm him. In minutes, he would meet his daughter for the first time, and he ached for it to go well.

The cloudless skies offered a false promise of a perfect, dry day but Dylan, highly attuned to the outdoors, knew it would rain in a few hours. He hoped his daughter would carry rain gear and boots. If she had asked him how he knew it would rain, he would have been unable to explain what he knew about the section of woods bordering Resurrection Bay on one side and the great Harding Icefield on the other.

When he received a terse email a week earlier, it was nearly impossible for him to comprehend. The sender informed him that Heather, a former lover and fiancée, had had a baby during the year they had been apart. Painfully, Dylan thought back to his last conversation with Heather. She had been trying to tell him something when an icefall ended her life near the summit of

Denali. Maybe she had been trying to tell him about Anja.

A daughter! The idea excited Dylan. He could form a deep relationship with her, show her important places, have conversations about life and go on dad-daughter outings. Anja would get to know Dylan's wife, Suzie, and love her like the mother who had perished on Denali. How Dylan longed for this daughter, even if she was in her twenties! Anja could be the child Dylan and Suzie never had.

In the email, Anja had explained that she was taking an administrative job in Anchorage with the Mountain Club regional offices. She'd be just a couple of hours from Dylan and Suzie.

His daughter's job impressed Dylan. The Mountain Club had started as a local climbing group in Seattle, and grown into one of the nation's most influential pro-environment groups.

It was probably most famous for its political lobbying, but it also organized outdoor events, published pro-environment books and raised public awareness of important issues. To get an administrative job with the big office in Anchorage meant she had talent.

Dylan had been in the local public library checking his email when he first heard from Anja. He'd looked her up and learned that her name was Anja Hart. Hart was her grandmother's maiden name and what her mother had gone by. He found out that she was only 28, but very successful in the non-profit environmental activist community.

What if she hated him? What if she blamed him for her mother's death and for not being a dad to her? What if she thought Dylan knew about her but didn't care to find her? What if she was coming to see him, not to form a relationship, but to unleash bitter resentment about a dad who never calmed her fears, kissed her goodnight or coached her soccer team?

As Dylan fretted about what could go wrong with the meeting, he let his eyes wander around the cabin.

On two walls, large windows let in the soft greenish light from the Sitka spruce forest and ushered in stunning views of Resurrection Bay and the range of snow-capped mountains surrounding it.

Another wall held Dylan's tools for guiding outdoor adventures, large gun safes, racks of bow-hunting apparatus and climbing gear. Near the ceilings, rusted and banned, antique leg-hold bear traps decorated the walls.

Dylan's eyes paused on a photograph of him and his journalist-wife Suzie. She looked so much younger than he. His salt-and-pepper beard and craggy lined face made him look like a mature movie action hero. In his mid-50s, Dylan's outdoor lifestyle had maintained his lean, athletic build.

Outside, some crows started making a ruckus. From their fuss, Dylan figured a hiker or hikers were probably coming through. Maybe it was Anja. Dylan felt awkward with people sometimes. He wished Suzie were home and not in Africa on a month-long assignment for Wild Outdoors. He didn't want to start off his relationship with Anja poorly.

A loud rapping at the door startled Dylan. It was probably Anja. *Should I hug her? Shake her hand? What if she cries? What if I cry?* Dylan decided to let Anja set the tone of their meeting.

When Dylan opened the door, he saw his daughter, looking like a blond Jennifer Lawrence. Her short hairstyle and androgynous outdoor clothes made Dylan wonder momentarily if she were gay. She had the athletic poise of her mother, but not the overt sweetness.

If he had planned it better, he wouldn't have been so stunned. His heart filled with affection for this beautiful person who would be his loving daughter.

"You Dylan?" the voice was deeper than her mother's and more confident.

Dylan gazed, speechless.

"Are you going to answer me?" the voice had a note of command in it. She was a top administrator for the Mountain Club. She definitely could command.

No hug. No handshake. Dylan found his voice. "Yes. Come in."

If Suzie had been at home, she would have noted that Anja jumped right into the purpose of her visit. She would have watched Anja coldly interview Dylan about her mother and any property her mother had left behind. Suzie would have noticed that Dylan stumbled over his words as emotional pain slowly welled up inside of him.

It was obvious that Anja was not the warm and fuzzy person Dylan hoped for. Not at all like her mother. Dylan wanted to ask her about her childhood, youth and college days. This wasn't going to happen.

"So you are an Alaskan now," Dylan found himself saying.

"Yes, imagine my surprise to learn my biological father is also here."

Dylan could not detect any warmth in her choice of words. "So you are a big shot in the Mountain Club?" Dylan tried to make conversation.

"I'm the chief fundraiser for all the Pacific Northwest. It's a good job."

While she talked, Dylan tried not to smile. He could see so much of Heather in this beautiful, competent young woman. He blurted out what he had wanted to say since she arrived.

"Anja, what kind of relationship do you see us having? I was hoping to attempt to make up some of the lost years."

"Lost years? They are gone. Right now I don't need anything from you, my absentee father."

Dylan felt a sting in her cold words. He steered the conversation to her mother, which is why she probably came to visit. When Dylan's stories started to peter out, Anja stood up.

"I'm holding open the possibility of continuing contact between us. You don't have a phone. I checked. Here's a smart phone." Anja placed a phone on the table. "Maybe we can be helpful to each other."

"You're giving me a cell phone? There's no reception up here."

Anja stood up abruptly. "Keep it anyway. I'll call you sometime."

Dylan looked at the cell phone like he had never seen one before.

"Once I get settled in Anchorage, I'll invite you up. Take the phone with you when you are in town. We can have another conversation."

As Anja strode toward the door, it was obvious that there would be no physical contact between her and Dylan. It was likely that she had some healing to do before she would ever call him Dad. But, at least she took the time to take the first step.

2

ANCHORAGE IS A DOWN TO EARTH TOWN with little tolerance for glitzy or incongruous luxury. A good meal is a good meal. It doesn't need to have crystal and silver to be excellent. Given this attitude, the Netsvetov Hotel stands out as an Alaskan anomaly. Taking up an entire city block, the Netsvetov shines like a polished diamond cast onto coarse black leather.

Built by oil-rich investors to rival the extravagant opulence of a fancy Dubai palace hotel, the people of Anchorage accept the existence of the Netsvetov with an uneasy tolerance that would be offered to a mad aunt living in an attic. Given Alaska's enormous oil industry wealth, the 500-room hotel is frequently booked up.

If the hotel is as brilliant as a diamond, the penthouse is the flawless, icy white refraction of light. Designed to appeal to the taste of a vain Saudi prince, with gold toilets, art museum paintings and flamboyant furnishings, only someone with an absurd amount of money to burn would ever rent the place. Colonel William Bolton had it booked for three months.

Sitting at an expansive Italian antique desk, wearing jeans and looking like a balding Clint Eastwood, business tycoon William Bolton made a call. About 150 miles away and seventy feet up a formation called High Rocks, the phone in Dylan's pack would soon ring.

Dylan placed a spring-loaded cam into a crack between two rocks. He was scouting a climbing route where he'd take his more advanced climbers. This climb featured a lower cave-like ledge big enough to park a minivan and a protected view of Resurrection Bay.

The High Rocks formation started with an intrusion of silica-rich magma, which cooled and hardened deep below the earth's surface. If Dylan had been a geologist, he would have known that High Rocks was an intrusive dike formed millions of years ago when quartz-dense magma filled a crack in the weak sedimentary rocks and hardened into a wall of stunning rigidity.

Dylan loved to climb it for the same reason shore birds nested in it, inherent stability. His hands on the rough, sparkling stone gave Dylan a feeling of confidence. He didn't know how he knew it, but somehow he could tell this formation's roots penetrated deep inside the earth.

High Rocks may have been a climber's dream, but graywacky or crumbly rotten rocks, surrounded it. No birds nested in the fragile, sandy area on either side of the sparkly gray High Rocks.

The phone ringing in his pack astonished Dylan; at first he didn't realize that the ring was coming from his pack. Dangling on a rope high above the big ledge and working the phone out of his pack, the screen glowed with a number Dylan did not recognize.

"Hello."

"Dylan Baker? Is this Dylan? This is Colonel William Bolton," the voice crackled.

What the hell? Colonel Bolton? Dylan hadn't heard from Bolton since he took a beating from some thugs who warned him to stay away from Heather, the rich man's daughter. Now, 28 years later, the old man's scratchy voice pulled Dylan back into the past.

"This is Dylan. How did you get this number?" Dylan eased the cam out of the rock. He needed to get down and concentrate on this call.

"Dylan, you need to come to Anchorage. We have to talk." Dylan could picture Bolton expecting everyone to instantly jump to his bidding.

"What's this about, Colonel Bolton? Is it Anja? Did she give you this number?" Dylan pushed the speakerphone button, shoved the phone into the neck of his climbing shirt so he could use both hands to climb down.

"Anja hates my guts. She thinks I treated her mother poorly, raised her wrong and that I throw too much carbon dioxide into the atmosphere. I got this number because I'm good at getting information."

"So that gives Anja and me another common interest, distrusting you." Dylan stood on a ledge and shook the rope to clear it from an outcropping.

"It doesn't matter what you think of me. You took my daughter away from me, and for that you owe me a few minutes." Bolton's words stung Dylan. Long ago he had started to shed his guilty feelings about Heather's death. Now this unwelcome caller was bringing it up again.

"I don't owe you anything. I rescued Heather from your sick family. Her death was an accident. Tell me why I should talk to you?"

"I don't want to talk about blame or any of your past mistakes. I want to talk about your carbon scrubber."

Dylan spun slowly on his rope above the ledge. *Scrubber? My college senior project? How did he know about that?*

"My carbon scrubber? Why are you interested in that?" Curiosity overcame Dylan as he slowly repelled downward.

"Why am I hearing there's interest in your carbon dioxide removal procedure, you ask? Did you not hear the rumors about a technology breakthrough?"

"I have no idea about rumors or why anyone would be interested in my old science experiment. That patent is expired now. Anyone can use that process. I have no interest in getting millions if it means I must leave my home and friends. Now you tell me, what are you doing in Alaska? Are you on vacation, or have you purchased the North Slope oil fields?"

"I'm here because a friend of mine, Greg Yeung, is running for governor, and he asked me to help. My interest in your carbon scrubber is private."

"That carbon scrubber was just an extra credit lab assignment. It will never be of any use because it doesn't work. The same biochemical reaction that removes CO_2 from a mixture of gases produces an alcohol that destroys the bacteria responsible for the reaction.

"My professor, who speculated that the culture produced the bacteria-killing alcohol, also told me there's no simple way to effectively remove the alcohol. I think you are just trying to annoy me because you found out Anja has come into my life." Dylan's feet touched the ground under the rock wall.

"So rumors about a breakthrough are news to you?" Bolton said.

"I've heard no rumors. I've heard nothing. The system is a curiosity, not a process to save the world. You can't make money on it. And I see no reason why we should meet. And now that I know you support Yeung, I'm voting for the other guy." Dylan terminated the call and wondered who Yeung's opponent might be.

3

THE MERCEDES-BENZ ZETROS is about the size of a large motor home and designed for military use with a 7.2-liter inline six-cylinder diesel engine that pumps out 960 lb-ft of torque through all six wheels. This monster off-road vehicle can go just about anywhere it wants and was equipped with a state of the art communication center, ATV/snowmobile storage, weapons locker and luxury furnishings. This particular Zetros had been built for a Mongolian millionaire who later changed his mind about the mansion-sized price tag.

The vehicle was parked out of sight near a newly renovated research station called Alaska Remote Climate Station or ARCS. ARCS looked like an old log-constructed national park building, but it was designed and built to study climate phenomena. The State of Alaska had repurposed the aging facility to study carbon dioxide and global warming.

Tango thought about the terrible road into ARCS and wondered how the state could keep it supplied. No wonder the ARCS was in such bad shape! Her Zetros had no problems with the snow and muddy, slushy journey from paved highway to ARCS.

The Zetros had just two semi-permanent occupants, the unit commander and a communications sergeant. However five additional guests were seated around a gorgeous teak dining table whose polished surface reflected the chandelier's light with a pleasing buttery glow. The guests appeared to be assembled from a major hotel chain, dressed in cook's white uniforms, custodian khakis, and building-engineer greens.

The leader of the group, and the only one dressed in white/grey winter combat camo was a fit looking, dark-haired, woman in her mid forties who went by *Tango*. She spoke with such authority, that the others had no problems following her orders.

"Memphis, has the ARCS security team finished their sweeps of the facility?" Tango's gaze fastened on a wiry, fit man wearing a green uniform. His nametag identified him as *Jeff* and his job as Maintenance.

"Yes, Ma'am. They are using cheap security equipment, like they don't expect any actual surveillance or attack."

"Good, then we can activate our micro cameras and mics." Tango turned her attention to a wide man in a cook's uniform. With his round head and big teeth, he looked to her something like a jack-o-lantern. "Salem, see that everything's up and working by tonight."

"Yes, Captain." Salem bobbed his large head.

"Reno, is the exterior surveillance set up as we planned?" Tango directed her comment to another man in a cook's uniform with *Josh* on his nametag. He looked like he belonged on a romance novel cover with his big chest and long reddish-blond hair.

"Yes, but it's hardly necessary. Everyone in there seems so intent on their lab and instruments, it's unlikely they'll want to come out in this weather." Outside a fierce gust of wind unsuccessfully attempted to rock the sturdy Zetros.

"Keep it on anyway. I know it's a pain to replace the batteries in this weather, but we want to know what they are up to when they are outside."

Tango glanced up at a clock on the wall. "When you go back to your quarters at the ARCS, remember you are to act like a contract support staff just doing your jobs. Show no overt curiosity about the work of the scientists. Unless they bring it up, don't discuss their work. Our surveillance devices should get us all the information we need." Tango gestured to a soldier sitting at a monitoring station, "Sergeant Denver will collect what we need."

Tango stood up, followed by a near simultaneous standing by the rest of the group. Tango smiled to herself, *These people are fanatics. They obey like no others I've ever worked with.* She delivered her last orders for the meeting.

"We have about an hour before any of you are back on duty in the ARCS, I want you to clean and check your weapons.

"Don't form emotional ties to any of the occupants of the ARCS. They are not enemies, but they are pawns that will probably need to be sacrificed for the greater cause. You need to be ready to kill, and soon." The fierce nods of her soldiers told her they would obey.

4

IT'S A FOUR-HOUR TRAIN RIDE from Seward to Anchorage. Dylan could have used Suzie's truck, but since Anja had invited him up to see her, and she was probably a proponent of public transportation, he wanted to do nothing to upset her. Dylan's hopeful side anticipated the beginnings of a strong father-daughter relationship, but he also had doubts. *I wonder why she invited me up.*

Dylan cycled his breath and centered his energies. His mind had a tendency to jump to several potential out-comes not all of which were positive.

An old friend, Peter Ivanov, picked him up at the Anchorage train station at 10:30 that night and let Dylan crash in the tent trailer set up in the driveway. Dylan went to sleep that night listening to raindrops on the trailer's roof. Peter had to get up early the next morning for work, or Dylan would have asked his friend to stay up late to get his advice on dealing with Anja.

THE NEXT MORNING, Dylan took a taxi to the building where the Mountain Club had their offices. The impressive size of the operation probably should not have surprised him. Where there's an enormous resource-extraction industry, there's the Mountain Club. The offices were on the top four floors of an elegant turn-of-the-century hotel that had been converted to offices and restored to platinum LEED certification by the U.S. Green Buildings Council.

The elevator opened onto a beautiful foyer with 19th century architectural ornamentation but poster-sized modern photos of some of Alaska's greatest natural treasures, as well as fascinating historical photos of Mountain Club greats.

To Dylan, the place looked like any other restored building of that era. He wondered what made it so green that it qualified for LEED certification.

The pretty young receptionist took his name and asked him to wait on a restored chair undoubtedly covered in recycled fabric. Dylan wondered if he should have worn used or recycled clothing instead of his new khakis and plaid dress shirt.

Soon a tall, Asian young man wearing used jeans, a fleece sweatshirt and a fancy ID tag that identified him as Eduardo, asked Dylan to follow him. What surprised Dylan, besides an Asian named Eduardo, was the security they had to pass through. It was what he would have expected had he been invited to the Pentagon's inner offices. Soon Dylan had his own electronic nametag, and found himself standing outside some large carved wooden doors.

Eduardo pushed a button and told Dylan to look up at a camera. Somewhere a buzzer sounded and the wooden doors made a loud click. Eduardo opened the doors to reveal a stunning corner office with a large glass and chrome desk. Looking comfortable behind the desk, Anja sat dressed in old jeans, a white dress shirt and a dark blue blazer. She wore no jewelry, but conveyed elegance.

"Do you need me for anything, Ma'am?" Eduardo asked her.

"Thanks, that will be all," Anja stood and came around the desk to offer a handshake to Dylan. He wanted a hug, but wasn't about to rush things. He shook her hand. It was cool and small but strong. The door closed behind as Eduardo left. Anja gestured to a white leather-and-chrome chair as she returned to her desk.

"What's up?" Dylan asked. *How lame is that?* he wondered.

"I wanted to make up for the abruptness of our first meeting. You wanted to form a relationship with me, and I was very uneasy with the idea." As she spoke, Anja played with her hair the same way her mother had done. It melted Dylan's heart.

"I would love to be a dad to you, but I understand if you want to take it slow. I'm very patient." Dylan sat back and tried to look non-threatening.

Anja leaned forward, "We should learn something about each other. Here's what I know about you. You quit college as a grad student in biochemistry and became a famous climber. You fucked my mom and left her, then became a reclusive hunting guide in a backwater Alaska town. Is that right?"

"I never knew Heather was pregnant nor did I find out I had a daughter until you contacted me. I loved her deeply. We were young. I was immature," Dylan felt a sadness creep into his mind. *Anja hates me.*

"That's it? You were *immature*?" Anja leaned back.

"When I lost your mother in a climbing accident, I went into a depression and found solace in the woods.

I've been here ever since." Dylan wondered if he should tell Anja everything about his woodland life—including a violent past that left him with barely controllable post-traumatic stress disorder symptoms.

"On our last visit, you didn't really get a chance to ask me any questions. OK, you lived off the grid not understanding that I existed, so I'm guessing you know nearly nothing about me," Anja said. "Now's your chance."

Curious, but not sure how personal to get, Dylan just said, "I didn't know anything about you until I got your email a month ago. Tell me about yourself. I'd like to know about your childhood since I missed the whole thing."

"Here's the short version of my life. Mother stashed me with my Aunt Louise when she went off to rekindle her relationship with you and climb Denali. You guys should have never broken up. When word of mother's death reached my grandfather, Colonel Bolton, he spared no expense getting custody of Baby Anja. He's a control freak.

"After he got custody of me, he showed no interest in parenting me, but hired nannies and boarding schools to rise me. He told me stories about what an asshole you are, and how I should avoid you. He told me that you did not want me."

"That's not true!" Dylan exploded out of his chair. "I knew nothing of you!" Just then a hidden door in the wall opened and two large bearded men wearing jeans and fleece sweatshirts burst out of the opening.

"Roger. Andrew. It's ok. There's nothing wrong." Anja gestured to the wall opening. The men eyed Dylan suspiciously and retreated. Anja looked at Dylan, "Security is pretty tight now. We got some threats from anonymous sources, so everyone's a suspect."

"I'll say security is tight. But I do want to be your dad. I would have loved to raise you. When I lost Heather, I thought all trace of her was gone." Dylan's naked pain was obvious to Anja, but she had no comfort to offer.

"In college, instead of business, I studied environmental science. It annoyed the Colonel. I finally found a real family in the environmental cause. At first, I was attracted to the most radical fringe elements of the movement. I firebombed bulldozers and chained myself to the gates of polluters. I soon realized destruction of polluters is neither a sustainable nor an effective way to cause change.

"I joined GreenWorld and helped haze whale hunters, but discovered I had a talent for raising money. After a while, I decided the real way to get change is to influence lawmakers and public opinion. Mountain Club recruiters found me and put me to work raising money."

"You worked for GreenWorld? Aren't they a radical group?" Dylan asked.

"Not radical. They believe in direct action to make change, and the general public is uncomfortable with that. I was in charge of courting donors. Turns out I am really good at asking for money." Anja gestured at the impressive office.

"Do you know how Bolton got my phone number?" Dylan asked out of the blue.

"He has the number of the phone I gave you?" Anja leaned forward. "That bastard! He finds out too much about me. I hate it. Did he call you?"

"Yes. Asked me about a carbon scrubber I worked on."

"Your carbon scrubber? You had a patent on that." Anja leveled her gaze at Dylan.

Dylan looked stunned. "How do you know about that?"

5

TANGO WANTED MEAT. She'd eaten only vegetarian fare since taking command of her small force, and with all of them now at the ARCS, she would take advantage of the time to do some hunting.

Studying the impressive weapons locker in the Zetros, Tango did not hesitate. She bypassed the most modern and expensive rifles, assault weapons, submachine guns and pistols to take out a wicked-looking crossbow.

She turned it over in the light admiring the elegant design and power of the weapon. The Ten Point Venom crossbow was custom painted in snow-camo, had a super light carbon fiber body and could deliver an arrow to target at 372 feet-per-second. Best of all, it had a quiet, light motor to recock, so Tango wouldn't need to struggle with a 185-pound pullback.

Her mysterious employer had not questioned the cost of anything for this mission. Tango never met her boss. They had only communicated through encrypted emails and voice-distorted phone calls, but Tango figured he must be either amazingly wealthy or work for the government. As she hiked through woods looking for a game corridor, Tango thought about the troops her boss had gathered.

As a top-notch mercenary, she'd worked with soldiers working purely for money and those working out of patriotism. But she'd never worked with any group like this. They were fanatically loyal to the boss who had hired Tango to lead the group. It was kind of eerie to have such amazing obedience from her troops.

Just then Tango spotted several caribou entering a clearing she had been watching. From the north, a massive bull with white antlers approached. He was going to have his way with a cow he'd been chasing. Too bad for him, Tango thought. She raised her crossbow. A caribou snort from her left made her pause.

From the south, a smaller bull with reddish/brown-colored antlers snorted again, ruining the romantic moment for the first bull.

Following that second snort, the smaller bull, aided by the slope of the land, closed the distance and slammed his antlers into the white-horned bull driving it back. The noise was louder than the firing range where Tango had worked on training her crew.

The volcanic rage and powerful clashes that followed that first charge were unlike anything Tango had ever seen. Tango figured that if the same amount of force were applied to the Zetros, it would rock dangerously.

Antler-to-antler, the bulls uprooted bushes, cut deep grooves in the forest floor and filled the air with steam. The peaceful clearing was soon a scene of destruction and flooded with a powerful mix of sexual desire, hate and fear. It soon appeared to Tango that the bulls' antlers were locked together.

The battle ended when the smaller bull lost the high ground. Physics prevailed. The larger, white-antlered bull got the smaller bull on a downhill slope. Viciously, the larger bull drove his rival downhill and into the javelin-like branches of a fallen tree. Sides heaving, the smaller bull seemed to lose consciousness as its wounds bled copiously onto the forest floor, its antlers still inextricably locked with the other caribou.

Tango watched as the larger bull struggled unsuccessfully to extract its antlers from his dying foe. Clearly, both animals were doomed.

Tango raised her powerful crossbow, took aim, and shot. The thrust of the bolt took it all the way through the big bull's chest and out the other side. One final cloud of breath and blood burst from the animal as its legs collapsed.

As Tango field-dressed the heavy carcass and took the parts she wanted, she reminded herself not to get tangled up in her current assignment. She hated working with ideologues; they would choose death over prudence.

Before she started her return hike, laden by 20 pounds of fresh meat, she paused to reload the crossbow. Tango would be ready for an animal attack, or the next order from The Boss.

6

"WE NEED SOME COFFEE." Anja stood. "Roger?" Anja called out in a normal voice. The side panel to the room opened instantly, as if Roger had been waiting for the call.

"Yes, Ma'am?" Roger eyed Dylan suspiciously. Dylan thought that dark-haired Roger looked as if he had played football in high school, but now was fighting a weight problem. "Could you send up some coffee and sandwiches? My meeting with Dylan is going to last through lunch."

"Yes, Ma'am." Roger cast a *you'd better behave* look to Dylan and left.

"Don't mind him. He's in charge of my security and takes his job very seriously."

"I should say he does. He gives me the creeps. Now tell me why you brought up my old patent. Bolton also mentioned it."

"Did he? Don't trust him. He hates you. He hates that you had sex with my mother. He thinks that if an adult woman has consensual sex, she's diminished. The truth is sex enhances an adult." Anja paused, scanning Dylan for a reaction.

Dylan, uncomfortable with Anja's frank talk about sex, wanted to change the subject. "Get back to the carbon reduction process. Why are you and Bolton bringing this up? It can't be a coincidence."

The large main door of room buzzed. Anja looked down at her desk to check a video display, then pushed a virtual button on her glass desk. A pretty young girl, hardly older than a teen, pushed a food trolley into the room followed by a scowling Roger.

"Thank you, Bridget," Anja smiled at the teen. "Roger, I want you to turn off the mics in this room. Dylan and I need to talk confidentially."

"Ma'am?" Roger looked as if she had asked him to jump into the Gulf of Alaska.

"Yes. Dylan is no threat. Take a break." Roger threw one more warning look at Dylan and sulked out.

Once the room was empty save for Dylan and his daughter, Anja motioned toward the lunch. "Help yourself. Hope you don't mind vegetarian." Striving for acceptance, Dylan reached for a piece of cheese.

Dylan just waited for Anja to tell him what she wanted him to know.

"The carbon scrubber. It's a long story. You know Governor Bates-Hardy?"

"No, I don't know him. I live in the woods. He lives in Juneau."

"Actually, no one lives in Juneau. The governor is a *she* and lives here, in Anchorage. But you know of our governor?"

Dylan nodded vaguely. *I should pay more attention to Alaska politics.*

"She's currently running for reelection against an anti-environment asshole named Greg Yeung.

"Bolton mentioned that he's campaigning for Yeung." Dylan grabbed a sandwich.

"That's expected. I'm certain Bolton figures his energy and financial services companies can benefit from relaxed environmental regs." Anja waved a carrot stick at Dylan. "He's curious about your carbon scrubber?"

Nodding affirmatively, Dylan said, "But why? That was an exercise for a biology class. I was supposed to design a device that would remove carbon dioxide. My design used a bacteria culture that eats carbon dioxide."

"So tell me, why didn't you make millions on it?"

"Easy. The process produced alcohol, which built up and poisoned my precious bacteria culture. There was really no way to extract the alcohol without an energy-consuming filtration process."

"If it was never going to work, why did you patent it?" Anja double dipped her carrot into some hummus.

"I was taking a personal law class at the time, and I did it for my law project. My protocol was original work and a new process. I never expected to make money on it. It was an assignment." Dylan picked up one of the sandwiches and eyed the brown contents suspiciously.

"What if I told you someone had solved the alcohol removal problem?"

"I wouldn't believe it. People way smarter and better equipped than me have tried it. The alcohol couldn't be removed without damaging the culture or requiring too much energy."

"Someone at University of Oregon did it funded by a grant from GreenWorld. A Professor Dave White, coupling your process with his team's research capabilities, identified a complex alcohol as the lethal agent and figured out a way to remove it. The result is a workable device that can scrub carbon dioxide from a passing airflow and convert it to an alcohol. The recovered alcohol can then be used as a gasoline additive."

"Interesting. I'd like to look at the research paper sometime. Anyway, what does this have to do with me? That patent expired years ago."

Anja turned toward the wall, "Roger, bring that box out here." Nothing happened. There was no Roger.

"You told him to turn off the mics," Dylan reminded her.

"Yah, but I was just wondering if he 'forgot' to do it." She pushed a virtual button on her glass desktop. "Roger, bring in that box from Oregon please."

The side door opened so rapidly, that it was as if Roger knew what she was going to ask for.

Dylan and Anja examined the long shiny aluminum box, which looked like something someone would send long-stem roses in but made of metal. Anja let Dylan open it and enjoyed his gasp.

"My god!" Dylan stared at the contents of the box. "The shape and simplicity suggests the work of a creative designer like Steve Jobs. It's so . . ."

"Elegant?" Anja said.

"Yes!" Dylan stared at a tiny model of a coal energy plant smokestack lined with valves and a sponge-like mesh filter. Dylan noticed Anja's grin. "My college prototype, built from miscellaneous lab items, was essentially a sealed flask modified with injection and venting ports. I injected CO_2 into liquid growth media containing soil bacteria and monitored the CO_2 levels in the gas that had bubbled through the media. This prototype is quite a jump from my murky bacteria solution in a flask."

"It's not a prototype, it's a model." Anja watched Dylan's face as he examined the highly developed application of his invention. "Dylan, are you an environmentalist?"

Dylan could sense powerful emotions behind the question. His daughter worked for the Mountain Club, but he thought an honest answer was best. "Like nearly every sport fisherman and hunter, I strongly favor protecting wildlife habitat. I feel a strong connection to the woods, the life existing in those woods and on earth. But, I do use fossil fuels and don't recycle as much as I should. You tell me, am I an environmentalist?"

Just for something to do, Dylan took a bite of the sandwich. It was actually pretty good.

"Your answer is good enough. Plus you are willing to eat tofu. I want you to meet with Governor Bates-Hardy."

Dylan stopped chewing. "Tofu? What? Me? The governor? Why would the governor want to meet with an outdoor guide from Seward?" Dylan's sandwich was forgotten.

"Frankly, I don't know why she wants to meet with you. It could have something to do with your bacterial process."

"How do you know the governor? You just moved here weeks ago."

"Remember what I do? I raise money. I also am on the board that decides where to spend money, political campaign money. Governor Bates-Hardy asked me to visit her as soon as I got here. Now she wants to meet with you."

"What does she want with me? Does she expect me to influence your campaign money decisions?"

"I really don't know why she wants to talk to you. But I would appreciate it if you did."

Dylan took another bite and chewed thoughtfully. With his mouth full, he nodded at Anja. *I wonder if I meet with the governor, if Anja will invite me back. If she just found out how much I want her in my life, she would let me be her dad.*

Dylan put down his sandwich, "OK. I'll meet with the governor. I hope she doesn't expect too much out of me."

7

DYLAN HAD BEEN WARNED that he would not be going to Juneau, but nothing could have prepared him for his visit to Governor Bates-Hardy's Anchorage mansion. The governor's grandfather, Ivan Hardy, was an energy baron who got into the market just before the development of pipelines from Prudhoe Bay. He built the imposing Victorian-style structure as a home and office building. The mansion truly dazzles visitors with its over-the-top ornamentation, turrets, stained glass and gaudy Victorian paint scheme.

Now more office building than home, the mansion served Governor Bates-Hardy as personal office space, campaign headquarters and an imposing venue for meetings and events.

An Alaska state trooper met Dylan's taxi at a large ornate gate. The trooper, a tall black man whose demeanor was all business in his dark blue shirt and smoky-the-bear hat, ran Dylan's ID through a scanner and inspected the taxi before waving it through.

Tucked behind the impressive facade of the main building was a nearly full parking lot and business entrance, which consisted of a low, modern-looking building that was recently added onto the mansion. As Dylan approached the entrance, two more troopers approached him. They appeared friendly, but conducted another visual search of Dylan and his ID in an efficient manner.

Once inside, Dylan passed through what looked like an airport security screening area with shoe removal and advanced imaging millimeter wave sweeping devices. Hopping on one foot as he replaced a boot, a man who looked like a very short secret service agent approached Dylan. The dark-suited man brought him a temporary photo ID.

"Mr. Baker, I'm Phil Small, your escort. This ID will fade to black within three hours, so you'll need to come back here for a replacement if you are here longer than that."

"I don't know how long I'll be here. I'm guessing just a few minutes." Dylan flipped the lanyard over his head.

"Follow me," Phil started up a set of stairs. "These were the carriage steps before the construction of the security building. The mansion was constructed long past carriage days, but Old Mr. Hardy wanted the place to look period perfect."

At the top of the stairs, they swept their IDs through a scanner and an amazingly tall, thick door opened into a large bustling room full of Reelect Bates-Hardy staffers and election paraphernalia. Over the din, Phil motioned Dylan to follow him through the busy worker bees of the Bates-Hardy campaign.

After negotiating a maze of hallways, small rooms and great halls, Phil stopped in an oval anteroom surrounded by antique chairs and heavy Victorian clutter. The room gave Dylan the feeling like he was inside a massive royal crown. "Mr. Baker, please wait here. The governor will see you soon."

With that, Phil Small walked to the anteroom entrance, stood blocking Dylan's retreat and assumed a don't-even-think-about-leaving stance with his feet apart and his arms crossed in front of his body.

Dylan sat in a lumpy Queen Anne chair and looked at the magazines scattered on an antique glass-topped table. Most were trade magazines about the petroleum industry and several general aviation glossies. Dylan opened one of the petroleum magazines and gazed at an article about hydrostatic pressure in oil wells. Not Dylan's type of reading. The aviation magazine he opened had an article about avionics that was also outside Dylan's normal reading material.

After 20 minutes of trying to figure out how the room had been constructed, Dylan heard a loud click in the tall doors leading to the main office. Phil Small gestured to the huge turret office and told Dylan to take a seat.

Visually, the room was full of design elements to impress a visitor, intricate tan and gold wood patterns on the floor, Oriental carpets, elaborate heavy carved furniture, swooping chocolate-brown drapes and an ornate chandelier. One wall was covered with gorgeous leather-bound books and brass sculptures of Alaska animals. Dylan wondered if the leather and dust smell came from the books.

Alone in the room, Dylan sat in an uncomfortably erect chair and waited while a wall clock ticked authoritatively near a large yellowed painting of a naked Rubenesque woman clinging to a Roman soldier. The room was so foreign; Dylan had to remind himself he was still in Alaska. He walked over to the painting to see if Rubens had signed it, when a side door suddenly opened and in strode Governor Leola Bates-Hardy holding a can of Dr. Pepper.

"Everyone who comes in here checks that painting's signature. It's unsigned, but it's real."

"I'm much impressed, Governor Bates-Hardy." Dylan found himself a bit tongue-tied standing in the grand room with the governor just feet away. She looked taller and more slender than her pictures, but still gave off the appearance of a competent executive in her fifties. He thought her casual haircut was a bit unnaturally dark. She wore perfect makeup, jeans, a red plaid shirt and an immaculate dark blue blazer.

"Call me Leola." Her eyes sparkled as she thrust her hand out toward Dylan.

"Yes, Governor," Dylan stammered as he shook her hand.

"Let's get out of here. I hate this room. It's called *the visitor's office*. My grandfather built it to impress visitors. My real office is through this door. I've got a surprise for you in there."

8

BOLTON SAT AT HIS MAHOGNONY desk and examined Jane's body as she sat before him and discussed their daily calendar. He liked her pixie-cut light brown hair and copper colored skin, but mostly her body fascinated and attracted him. Sometimes she looked like a 12-year-old girl with her small stature and narrow hips. Sometimes she looked like a teenaged boy on the verge of sexual maturity with her soft cheeks and slight athletic body.

Bolton knew Jane's father was a blond Norwegian who had met Jane's Vietnamese mother at work. The result was a small, beautifully formed mixed-race woman who was emerging to be the focal point for his forbidden sexual fantasies.

Jane dressed as if she had read a book about modern female executive attire, charcoal-gray blazer and slacks, a white silk top and flamboyant gold jewelry. Although Jane dressed and presented herself at all times as a thoroughly professional political campaign manager, she tried to overcome her youthful appearance by using heavy eye makeup and gaudy bracelets and a subtle neck tattoo, a small butterfly just under her collar. She knew the look didn't work.

Before taking this job, she had tried to make herself appear more mature by cutting her hair shorter, but it just made her look more like a young boy instead of a girl.

Bolton wished she were dying to submit to his wishes. But something about the way she flinched when he let his hand rest too long on her back told him it wouldn't happen.

It wasn't just her body that interested Bolton. Jane's resume made her perfect for his needs. Her academic record hinted at an IQ in the stratosphere, and she had run a successful highly contested Texas state senate campaign. She had also helped squeeze a Republican governor into a solid blue state through the use of a well-run Super PAC: a 501(c)(4) nonprofit that can raise and spend unlimited amounts of secret money from any source.

Bolton had hired her due to her quick, strategic political mind. Even if Bolton couldn't get her into her bed, he would have his way with her mind.

As she read down the list of what they would be doing for the Alaskans-for-Progress campaign that day, Bolton imagined her sitting on his lap and nuzzling into his chest.

"Colonel Bolton, are you listening to me?" Jane's big brown eyes looked up at him from across his massive desk. He'd need to get a smaller desk so she'd be closer. He loved hearing her soft Texas accent.

"Why yes, Jane. You were telling me about my presentation at the Energy Forum at the University of Alaska." Bolton actually looked down at his notes.

"I want you to pay particular attention to the benefits Alaskans get from the energy industry and the amount of money spent on safety and cleanup. Since the 1989 Exxon Valdez spill, billions have been spent by the energy industry and 13 wildlife populations have returned to normal. Let's hope your debate opponent won't mention that 19 populations are still struggling."

"But improving?" Bolton asked.

"Some are, but the most charismatic mega fauna species are not doing well. The orcas in that area for instance, have not produced a single calf in 25 years." Jane made a note on her brief for Bolton's debate.

"Those damned environmentalists. They look for a cloud near every rainbow." Bolton couldn't understand why people didn't realize that God provided resources for man to use. As soon as the planet became unfit for humans, God would return and call true believers to heaven. In a way, using these resources moved up God's timeline. How the Colonel loved thinking he could influence God's actions!

"You should mention The Permanent Fund dividend to your audience. In past years it was nearly $3,000 per resident. Be sure to express your confidence that the price of oil will zoom back up." Jane moved her notes in front of her chest. Bolton made eye contact.

"How brilliant that the Alaska petroleum industry has made the entire country cheer for profits. If only something like this could be done in California."

Bolton often daydreamed that he could get offshore drilling regulations loosened up in Southern California. Californians were afraid of spills. *Didn't they realize that oily beaches are just part of nature? In fact, oil spills create jobs. If there was an accidental oil spill, why not hand out paper towels to clean the coastline rocks to minimize use of cleaning chemicals?*

"Colonel, my source in the governor's campaign has leaked to us that Governor Bates-Hardy is going to propose new drilling regulations along the Arctic coastlines." Jane smiled at this nugget. "She's also going to bring up the effects of climate change on Alaska and how residents should prepare for it."

"Climate change? That's perfect. It will mean more regulations and limits on resource extraction. I can't believe our luck! Alaskans will never vote for someone spouting that crap." The Colonel rose and looked out his penthouse window at Cook Inlet and the vast snow-capped mountains in the distance. "Now I see why you showed me how the Permanent Fund will benefit from allowing more drilling on coastal and state lands. You are a keeper."

Jane smiled at the compliment. "Be sure to point out that the average resident could see record dividend checks in two years if Greg Yeung is elected, and how the Alaska state income tax could become permanent if Governor Bates-Hardy is reelected. Voters should see Yeung versus Bates-Hardy as a pocket-book issue."

Bolton moved to Jane's side of the desk and stood a bit too near her, his belt buckle at her eye level. Jane turned away and stood up and moved toward a floor-to-ceiling window while looking at her tablet.

"You have a meeting with Mr. Yeung this afternoon. He's campaigning at a fishing derby in Homer, but his floatplane should get him back here by late this afternoon. Since the money you are giving his campaign is in the form of a 501(c)(4) organization, there can be no coordination between the money you are spending on issue ads and what Mr. Yeung's election committee is planning."

"Yeah. Yeah. That's old news. We never talk about coordination." Bolton winked at Jane. "But he and I can discuss fishing, right?"

"Right. You can tell him about your debate at the University Forum since it's not an overt campaign event. You are just an outside energy expert discussing the future of Alaska. I've sent copies of your remarks to the major Alaska news organizations, so they should all be there. And make notes of your conversations with Yeung in case the feds decide to look at the Super PAC."

"Jane, you are the best. I could just kiss you." He noticed Jane moved to put the desk between them.

Jane quickly angled toward the door and its access to her office. In her departing glance, she noticed that Bolton had run his finger down a tattered yet laminated card taped to his desk. Jane knew that card; it was Bolton's numerology guide. Whenever he had an important meeting, he would calculate the number associated with the person, the meeting location or the date.

He used the characteristics of that number to help him select the best date or location. She knew he would soon call his secretary, a plain Midwestern matron named Elizabeth, to look up the birthdates of the others at the meeting so he could put the information into his calculations.

Shaking her head as she crossed the reception area to her office, Jane considered the fact that numerology, like astrology, has been around for millennia and used by people in all walks of life to make daily decisions.

But Bolton had taken lucky numbers to another level. He believed that the number derived from people's names or dates provided insight into behavior or timing. For example, Bolton had read that the number "1" represented *creativity,* while the number "8" represented *wisdom* and *financial security.* Bolton often selected a meeting date based on the fact that the digits summed to an "8".

If I'm going to optimize my working relationship with Bolton, I think I'll pay more attention to his decision-making processes, Jane thought.

9

DYLAN FOLLOWED THE GOVERNOR through tall, imposing doors and instantly found himself in a comfortable and functional office with modern sensible furniture, mini bar and fireplace with a wood fire crackling quietly.

"I'm trying to quit Dr. Pepper. Want some coffee?" She went right to the mini bar and poured some pale brown coffee beans into a conical burr grinder then tried to talk over it. "Thanks for coming over. I know Seward is not exactly close by. Did you say you want coffee?"

"Yes Ma'am. Coffee sounds great." Dylan stood awkwardly not knowing if he should sit down or stand by the sink and watch the governor of Alaska make him a cup of coffee.

"I get these beans from a roaster in Seattle. Don't tell any of my Alaska friends in the coffee business." She poured the grounds into a simple french press and added hot water from a dispenser. "I suppose you are wondering why I invited you here when we are so wrapped up in my reelection activities."

"I can't figure that out. I mostly live in the woods with my wife and hang out in a very small town. I can't really do much to help you turn out the vote, and I don't have much to contribute to your campaign but my vote." Dylan made a mental note to register to vote as he watched her stir the coffee mixture with a well-used wooden spoon.

"I really want your vote," a small smile lit up the governor's face, "but that's not why I invited you here." Slowly the governor pressed the plunger down on the coffee maker. "Do you take cream? If so, please taste this first. I think you'll like it the way it is."

"I drink my coffee black, mostly out of necessity. It's hard to get milk out to my house. It's pretty remote." Dylan watched her pour the coffee into two handsome mugs.

"How does your wife handle that?"

"Suzie is a good sport, and she knew what she was getting into when she married me." Dylan blew across his mug.

"Your wife, she's a journalist, right?" The governor strode over by the fire. "Let's sit down here. I'm tired of acting like the leader of the biggest state in the union."

Dylan followed her wondering how she knew so much about him. When she gracefully sat down on the rug by the fire, Dylan joined her on the floor so that both of them were cradling their mugs and leaning against some sturdy leather sofas. He suddenly felt comfortable with this woman.

"So, what do you think of the coffee?" Leola Bates-Hardy looked at him expectantly. Her eagerness to know how he liked her coffee completely disarmed him.

Dylan blew across the fragrant brew and took a small, noisy sip. "This is delicious." Dylan had never tasted coffee so rich and deep. "Is there a bit of chocolate in this?" He took another sip.

"Nope. It's just coffee, but I like it for the finish. You can nearly taste some floral notes in it."

Dylan doubted he would taste floral notes, but he really liked the coffee. "Why would you ever drink Dr. Pepper when you can have this?"

"I know. I drink pop when I'm on the run, and it's been crazy around here. Back to why I invited you here. Can you guess why?"

"The invitation came through my daughter, so maybe it has something to do with her?" Despite the governor's ability to put him at ease, Dylan knew that this person was a powerful and effective politician. He did not want to say anything that would cause complications for Anja just when he felt close to becoming her dad.

"You get an A for guessing. Anja holds sway over a large political fund and, more importantly, has the ear of hundreds of potential volunteer campaign workers. Greg Yeung, my opponent seems to have nearly endless funds to spend, and he's hurting us with his constant attack ads."

Dylan suddenly lost interest in the coffee. *Was she going to ask him to influence Anja?*

"But I asked you here today because of what's in that box." The governor gestured with her mug to the modern executive desk in the corner. On top was a metal box of the size and shape to hold a dozen roses.

10

JANE KOSS HAD FINISHED RUNNING the staff meeting for Colonel Bolton's Super PAC and sent her workers off to their assignments. It was time for her to enjoy some time alone in her office.

Although Bolton's 501(c)(4) was a nonprofit that she had organized as a social welfare organization, everyone knew it really was a mechanism for getting secret, unregulated and untaxed campaign cash to a candidate or cause. All she had to do was to make every political ad end with, "call Governor Bates-Hardy and tell her to stop destroying Alaska's economy" and voila! An expensive political ad suddenly became protected free speech—even if secret foreign or out-of-state donors provided the money.

When Jane had entered South Texas School of Law as a scholarship student, she could not believe her escape from poverty. Her father had worked in an office at a poultry processing plant to pay the bills. Her mother, a Vietnamese refugee, had died in childbirth having her fifth child and only daughter, Jane. Now Jane hobnobbed with millionaires and ran a Super PAC with over $60 million in potential assets. Some of the donations were promises, but Bolton told her to fully expect the money to be deposited.

Jane's dream was to become the go-to person for Republicans seeking national office, maybe even run a presidential campaign someday. But first, she had to build a resume. This meant she needed more experience with a 501(c)(4) organization. When Colonel Bolton had asked her to run his, it seemed a dream come true. A chance to get a small government candidate into office would align perfectly with her goals and political beliefs.

As Jane got to know Bolton and his reasons for getting Greg Yeung into the governor's office, she began to have second thoughts about her work. First off, Bolton was a dick. He seemed like an intensely intelligent man with a middle school emotional balance.

The second doubt came when she accidentally uncovered the source of some of Bolton's contributions to the 501(c)(4). He had not reported loans against some of his companies to stockholders thereby unlawfully boosting the stock price. This was fraud, and he could actually go to jail if he got caught. His head accountant, a jittery fellow who always wore a bowtie, had helped him pull off the job and earned the honorary title of Chief Financial Officer, but now seemed as nervous as a crooked CFO could be.

Bolton's unrelenting self-confidence seemed to be deserved, because a random bump in the stock market allowed him to pay off every debt before anyone got wind of what he was doing. The guy was a truly amazing financier and industrialist. He could get things done. And once Greg Yeung was in office to clear out environmental regulations, Bolton's huge gamble on Alaska mineral rights would pay off to an unbelievable degree. His wealth could approach that of the 10 richest people on the planet.

When Jane learned of Bolton's lie to his stockholders, she intended to quit. However, Bolton was persuasive. He promised that his lie to stockholders was the one and only time he would commit fraud, it was for a good cause and nobody got hurt. He was able to convince her to stay on as the executive director of his 501(c)(4).

He explained to her that his fear of prison bordered on a near phobia due to the number values of letters associated with a prison sentence. She believed him. He seemed genuinely terrified about being locked up. She knew she could ignore his adolescent sexual overtures if she could get his recommendations for the next campaign job that would come up.

She looked out of her office window to admire the quaintness of Anchorage, the deep greens of the forests accented by white cloud wisps and pale blue-greens of Cooks Inlet's waters. She let her gaze travel to the vast beauty of the snowcapped mountains in the distance. Living in Alaska for the last few months and learning to love its pristine vistas, she had also begun to wonder if Alaska would truly be a better place with fewer environmental regulations.

11

DYLAN LOOKED AT THE ALUMINUM BOX on the governor's desk. "Is that . . ."?

"Your carbon dioxide removal system applied to real life? Yes. Did Anja show it to you already?" The governor made no move to stand, so Dylan remained on the rug with his coffee.

"I'm confused, Governor. My original design didn't work. I mean, it would remove some carbon dioxide from the air, but the bacteria quickly killed themselves in their own wastes. It wasn't feasible."

"We'll talk about that shortly but first, you need to know a secret about me." Leola lowered her voice. "I'm turning into an environmentalist."

This revelation confused Dylan. Generally Alaskans were skeptical of environmentalists, who were typically outsiders coming in to tell rugged, independent Alaskans what to do. If Governor Bates-Hardy admitted to her constituents that she was pro-environment, it would be like saying she's "anti-jobs", or "pro-government meddling in the affairs of private citizens".

"You can't tell anyone yet. I need to get my ducks in a row first."

"But won't this play into Greg Yeung's advertisements? He is trying to brand you as pro-regulation and anti-freedom."

Leola stood up. "Come over here and look at this." She strode over to her desk and pushed aside the aluminum box. She opened a laptop and brought up a document.

"A mayor of one of our most northern towns sent me this report. He says that, due to warming, the tundra lakes are disappearing, along with drinking water and habitat for birds. Hunters go out for moose, but because of the silt from eroding riverbanks, their boats get stuck. They need a boat to get their meat from the field to their homes.

"Our scientists have documented that the Arctic is warming twice as fast as the rest of the planet." The governor tapped some laptop keys and brought up a report from a state commissioner. "This report lists some of effects of climate change we are already dealing with, some massive coastal erosion, increased storm affects, sea ice retreat and permafrost melt. You know what happens when permafrost melts? Methane!"

"Methane?" Dylan put down his cup.

"Yes, methane. As a greenhouse gas, it's 25 times worse than CO_2. This is urgent. We need to stop climate change."

Dylan had lost interest in his coffee as he looked at the alarming temperature and sea ice graphs. "Are you confident that the Arctic warming data does not just represent normal fluctuations that happen whether or not humans are on the planet? Your statements are pretty alarming."

"Yes. I'm certain that humanity is significantly contributing and 98% of geoscientists agree with me. And I hate to say it, but it's actually worse than most people believe. We've already started the relocation of several coastal villages and this will increase ten-fold within the next five years."

Dylan wondered why she was telling him all this, but he felt he had to tell her where he stood on the issue of climate change. "Governor Bates-Hardy, you know I'm a hunting guide. Overall, hunters are a pretty redneck group, but we love fishing and hunting. As a result, we cherish clean water, air and wildlife habitat.

"Unlike a minority of commercial fishermen, we hate overfishing and overhunting. We sportsmen value wildlife regulations. When someone gets busted for hunting out of season, we celebrate." Dylan looked down at his nearly empty coffee cup.

"Should I make another cup for us? My cup is empty too, and I'm thinking about a Dr. Pepper." Leola closed her laptop and returned to the mini bar.

Dylan stood by while the governor brewed another amazing cup of coffee for them. "Why are you telling me all this? Shouldn't this be something you share with voters? Can't you trust them to make a good choice?"

"The voters don't always make good choices. They elected Donald Trump a while back. They can be manipulated if they are not paying attention. I'm going to the voters with all of this, but I want to pick the right circum-stances. We'll need a huge earthquake of a change to get the voters' attention."

"And what might that be?"

Leola pointed to the aluminum box. "It's in there. It's your CO_2 scrubber."

12

COLONEL BOLTON WALKED to the meeting place, Margaret Eagan Sullivan Park just a few minutes from his hotel. Despite the wind and rain, there were still families walking their dogs and kids playing. Bolton wondered what kind of people these Alaskans were that they didn't just play inside during bad weather.

He did what he was supposed to do. On the hour, he dropped a newspaper into the trash, got a drink from the public fountain then retrieved the paper. He didn't know how it was done, but mixed in the soggy paper was a slip that told him to walk one block north and look for a black 4-door Jeep Wrangler. He looked around Sullivan Park and saw two of these vehicles in the parking lot, but different colors. Bolton hoped there would be only one black vehicle a block away.

The Colonel found a black 4-door Jeep and opened the passenger door. Inwardly he groaned, *a fucking Arab at the wheel. This better be worth my time.*

The driver, a short, dark-skinned man with a squarish face and small black moustache, held out his hand. Bolton thought he looked as if he were dressed for a funeral. "Colonel Bolton, I recognize you by your pictures. I'm Ayushmann Krishnamurthy, you can call me Krish."

"OK, Krish. Call me Sir or Colonel Bolton." the Colonel extended his hand. "Are you representing the Saudis?"

"Oh no, Sir. I'm not Saudi. My family is from India. I'm a businessman and broker. My job is to bring various parties with common interests together." Krish held out an envelope. "Here are the details of the wire transfers scheduled to go into your 501(c)(4)."

Bolton, annoyed by Krish's high-pitched voice and foreign accent, took the envelope, opened it and scrutinized the contents. "This is more than we talked about, but why the delay in the transfers? We have a cash flow problem, and we need the money now."

"Oh Colonel Bolton, we can't adjust the schedule, but there is very much more than that available." said Krish in his singsong Indian accent. "The Saudis, Iranians, Nigerians, Russians and Venezuelans are very interested in your proposal. They want to meet with you."

"They should be interested. Together we can enhance our mutual interests. It will be like the OPEC cartels of the 1970s. Those were the days!" Bolton licked his lips.

"Yes, Colonel. But it was very, very difficult to get these people together for that meeting you suggested. I need some assurances from you now for this meeting to proceed." Krish looked a bit nervous saying this to Bolton.

"Assurances? What more can I give?"

"They want to meet at your hotel office suite after it's been swept for surveillance devices," said Krish.

"Got it. I have it swept regularly anyway. Everyone from our political opponents to the feds are very interested in our 501(c)(4)."

Krish looked nervous. "They also want you checked before the meeting, and they will also submit to a personal security sweep."

"No problem. I would have insisted on that anyway."

"One more thing. Only you can be at the meeting. No staff members or aides are permitted."

This caught Bolton by surprise. "I thought that Saudi prince what's-his-name never went anywhere without bodyguards, aides and some of his wives?"

"Oh, they will all be there, but outside your office or in rooms downstairs. Are these conditions approved by you?" Krish looked Bolton right in the eye.

"Listen, Krish. My goals and the goals of the others are perfectly aligned. We all want the same thing, but it's not going to happen unless we cooperate. If we can put Yeung into the governor's mansion, we'll be able to control 60% of the Arctic's oil reserves." Bolton's voice was getting louder.

"Not only that, but once the other producers see what we are able to do, a bunch of them will hop on board." Little drops of spit were spraying out of Bolton's mouth. "Once that happens, we'll have the power to make the political and social changes that we want."

Krish cleaned his glasses and leaned away from Bolton. "Yes. That's what they are thinking, but a lack of secrecy could cause everything to unravel."

"Very well, you tell them that I agree. You tell them that I can keep a secret, and I expect them to also. Tell them to move up the schedule of deposits. We are on fumes right now with what we're spending on attack ads."

Bolton concluded the meeting by shoving the envelope into his raincoat pocket and exiting the Jeep. He walked away as rain pelted the Jeep. He appeared composed, but inwardly he was jumping up and down in excitement. This is going to work. It's really going to work.

If Bolton had been more self-aware, he could have heard a tiny voice deep inside telling him to be cautious.

13

DYLAN AND LEOLA CARRIED their steaming mugs back to the desk with the boxed up CO_2 scrubber model. "Don't you want to see it?"

The box appeared to be similar to the design that Anja had shown him.

She set her mug down and opened the box. Leola invited Dylan to pick up the model.

"Anja did show me a model. It looked similar to this." Dylan said.

"Actually Dylan, this is a more advanced model of the carbon-dioxide removal system that Anja showed you. The actual prototype is affectionately called *the Scrubber*."

"This is amazing." Dylan spoke in a reverent voice. "The design has evolved a lot."

Dylan examined a tube that had been formed into an "L" shape. The horizontal, short end of the "L" sprouted injection ports and tiny ducts that emptied into a small receptacle. The longer, vertical part of the tube was studded with venting and sensing ports and was filled with some kind of 3D mesh that almost looked like sponge cake.

"You have a now-expired patent on the biological process part of this device. The University of Oregon holds the rights to the hardware and auxiliary systems."

She pointed to small tubes leading from a little box to cooling coils embedded in the sponge cake. "Coolant is circulated through these coils to prevent heat damage to the living bacteria."

"What's this part?" Dylan pointed to a tube leading from the sponge cake to a metal bottle.

"That's a collection tank for the alcohol. It can be sold as a fuel additive for cars."

Dylan appeared visibly moved as he held the model. "Eventually, future versions of this device could go into exhaust pipes, submarines and space ships." He looked up at her, "This technology could change the world."

Dylan traced the direction of the flow of exhaust gases through the model. "This is the earthquake that would get the voters' attention, isn't it?"

The governor said nothing, but smiled at him.

Dylan looked directly at her. "You want me to be the spokesperson for this technology, don't you? An Alaska woodsman who both participated in the original work and supports this new technology would indeed catch the public's attention. It will get you elected if you declare a war on carbon and have the right weapons for the fight. This will change coal-fired power plants from being the most polluting to producing nearly no CO_2 pollution."

Leola picked up her mug and gestured to Dylan, "Follow me. I want to show you something." Dylan put down the model and followed her back into the dusty Victorian office where a painting hung on a dark wall. The painting was of a slightly sepia 19th century stern-wheeler being loaded at a river dock.

"What do you notice about this picture?" The governor stood by and watched Dylan.

"It's an old painting of the Yukon Queen," Dylan said as he examined the name of the ship in the picture. "It's being loaded at a dock, probably on the Yukon River."

"That's right so far. What are they loading onto it?"

"The cargo looks like what you'd expect for the 1800s, furs, ore, timber." Dylan looked closer, "and livestock."

"That's the cargo, but what's the fuel?" The governor pointed at the painting.

"Coal. They are loading coal from the dock. But where did they get it? Was there coal mining on the Yukon?"

"Yes, but it was rotten coal. It burned very dirty and caused all kinds of problems with engines and even the crews. The coal was full of sulfur dioxide, mercury and other pollutants. That's why the smoke stacks look yellow inside."

"Why are you telling me this? Is the scrubber not so useful in coal-fired power plants?"

"I'm telling you that a huge discovery was made up by where the Yukon River flows near Denali. Your carbon scrubber is going to be used in a test up there to see just how useful this device really is."

"What kind of discovery? Something to do with dirty coal?" Dylan eyed the yellow smoke stacks of the stern-wheeler.

"You'll find out when you get there. It's a newly repurposed remote climate station, and I want you to observe the tests and report back directly to me and to Anja. She's on my butt, and I need her support, not her kicks."

"You want me to go to the Denali area? I'm not a scientist. That carbon scrubber invention was just a classroom project that I never expected to become practical. I doubt that it's going to be the earthquake you need, and I don't know how my attending an experiment will change any of that." Dylan looked as if he wanted to leave.

"It's going to be an earthquake when you say you saw your carbon scrubber process work and that you believe in the technology. It will assure your place as a hero of Alaska and my reelection."

"I don't know about the hero part, but I'd hate to disappoint my daughter or see that Greg Yeung character elected. If I decide to go, when would I leave?" Dylan set down his mug.

"Right now, there's a car waiting."

14

FROM SOMEWHERE, Phil Small came into view and pointed to Dylan's ID badge, which was turning gray. Phil gave a *follow me* gesture and whisked Dylan back out of the mansion into the parking lot.

Dylan didn't even have a chance to say goodbye to the governor, and he didn't remember agreeing to go on this trip. Only the idea that he and his daughter could become a family allowed him to accede to the governor's wishes.

Phil opened the door to an Alaska Department of Natural Resources pickup to see a young woman of Native descent wearing park greens and an Alaska State Park Ranger baseball cap. "Hi, I'm your ride out to the airport."

Dylan looked confused. "I need to go home and pack a bag if I'm going to a remote research station."

"No problem, Mr. Baker. Deputy Director Small delivered your bag an hour ago." She indicated a large, unfamiliar duffel in the back seat of the truck. "We're set to go, if the weather cooperates."

"Wait. That's not my bag, and I need to go home."

The ranger looked confused by his outburst. "Why certainly, Mr. Baker, I guess I got my instructions mixed up. Deputy Director Small had me get your duffel from your daughter. She told me to bring it to you for your trip."

So Anja really wanted him to go on this trip. Dylan looked at the duffel. Anja must have had one of her boys purchase some clothes. Maybe, if he did this for Anja, she'd continue to warm up to him. "How long am I going to be gone? I have things to do in Seward." This was actually Dylan's slow season, but he didn't want the ranger to think that.

When the park ranger smiled, her merry eyes were nearly hidden in her cute, wide face. "I have no idea. I just got word to pick up your duffel and drive you to the airport. Would you like a number for the governor's staff?" She dug around for her phone.

"No. I have Small's card somewhere." Dylan became quiet on the trip to the airport. It made him feel uneasy to live in a world where everything happened so suddenly.

This morning he visited his daughter's office for the first time and found out she had a difficult childhood. Then he had a private meeting with the governor to learn his college science project could change the world, and now he was flying to somewhere remote without even getting a chance to know what was in his duffel. He hoped Anja had packed the right gear.

With all these things going through Dylan's mind, the fact that he hadn't been there to protect and guide his daughter caused him the most concern. Guilt surged through his consciousness. He tried to tell himself that Heather should have involved him in her pregnancy. But in those days he had been immature and selfish.

Maybe Heather thought he would not handle the information about their child in an adult way. Dylan wished he could change the past. At that moment, Dylan longed to climb a rock cliff in Mexico or dig a snow cave on Mt. Alice. He wanted some control and time outside by himself.

The ranger drove to the part of the airport where mostly small charter planes lined up on the apron like hopeful taxi drivers. After parking in a reserved parking spot, they walked through gusty winds and large raindrops. They passed several turboprop planes and stopped at an elderly Beechcraft Bonanza 35 with a V-shaped tail.

"Wait. This is my plane? Aren't these V-tails called doctor killers?" Dylan looked at the unconventional V-shaped tail.

From behind Dylan a voice startled him, "These were called the doctor killers because they are tricky for amateur pilots to fly. They begin to come apart when a pilot takes them past their top speed, typically when flying doctors to rural emergencies. These planes can fall into a spin when the controls are treated roughly."

Dylan turned around to see a slender, almost spidery-looking, older man with crooked teeth, a friendly smile, and a blue flannel shirt. His well-used dark blue woolen baseball cap said Flapjack's Alaska Charter Flights in gold letters, and his worn leather jacket had a battered nametag reading Flapjack Curtiss. Just then Dylan noticed the park ranger had left and was getting into her truck. She waved to him and turned away.

"This your duffel?" Flapjack bent down to pick it up. "Pretty fancy duffel. Heavy too. Good thing you are the only passenger on this trip. Let's get it into the plane before this rain really starts to come down." He grunted as he pushed the bag into the passenger back seat.

Dylan started to enter the plane when Flapjack called out, "Smile."

Dylan turned to see Flapjack holding a tiny camera whose flash made Dylan blink.

"I got a picture of every passenger I've ever flown," Flapjack pushed the camera into his pocket.

Once in the Bonanza, Dylan noticed the old plane was in amazing condition, as if it had been restored for an air show or museum. "So you think this plane is plenty safe?"

"I wouldn't fly anything else. It's got 285 horse-power, retractable landing gear, cruising speed of 230 mph, a range of nearly 900 miles and a 20,000-foot ceiling. I've been flying this bird since 1998."

"Where are we going?" Dylan loved the immaculate retro interior of the Beechcraft and the Easter egg pastels used to paint the exterior.

"Don't know. I was told to file a flight plan for Nome but you'll get a satellite phone call once we are airborne with a new destination." Flapjack pointed to a satellite phone on the seat behind the pilot. He completed the preflight check as Dylan toyed with the phone.

Many towns in Alaska have only one way in or out, small planes. It's common for a village to have river access several months of the year, but travel beyond hunting and fishing grounds usually involved using a small plane flown by a bush pilot.

Dylan knew that the Beechcraft was not a bush plane since most are tail-draggers that have fixed wheels, which can be fitted with skis, floats, or huge bouncy tundra wheels. He was sure that Flapjack would hate to land his pretty Bonanza on a gravel runway and risk a rock chip on his perfect paint job.

Once preflight was done, Flapjack took the pilot's seat and started pumping a lever near his seat. "Gotta build up some pressure in the fuel system," he said apologetically to Dylan. "Pressure's up," Flapjack announced.

"Clear the prop!" he yelled out the open cockpit door, then pressed the starter button. With an asthmatic wheeze, the engine turned over and over, but didn't start. Just as Dylan thought the wheezing was starting to slow down, the engine coughed. With a puff of smoke, the prop started spinning rapidly and the engine roared athletically.

Dylan watched Flapjack manipulate levers and tap on the oil pressure gage. Once he was satisfied the plane was ready, he called the tower for instructions.

A few minutes later, as Anchorage and the more populated parts of Alaska faded into the background, Dylan enjoyed the stunning views of mountains, waterways and forests below the Bonanza. The phone in Dylan's lap chirped, startling him.

"Answer it," Flapjack called out. "Get the new heading and enter it into the navigation screen. "

Dylan answered the sat phone. He was pretty sure he was talking to Phil Small, but Phil never identified himself.

After Dylan entered the new destination, Flap-jack scowled.

"There's some bad weather over there. Call him back and ask what he wants me to do. I'd hate to take this pretty Bonanza 35 V-tail past its safety envelope. You're not a doctor, are you?" Flapjack pointed to some very dark clouds far to the northwest.

"Dylan called back the last number, but there was no answer. He left a message saying that he might be delayed due to weather, and that his pilot would find somewhere to sit out the storm rushing in."

Dylan thought about the V-tailed plane he was in and really hoped the storm wouldn't be too bad.

15

TANGO SAT AT THE POLISHED TEAK table in the Zetros, and finished cleaning and reassembling her Sig Sauer MPX-SD. As she screwed on the silencer, the smell of gun oil and a faint whiff of gunpowder made her feel content. She thought about her amazing sub-machine gun. It fired the .357 pistol ammo she liked at full auto, but the silencer suppressed the sound to an astonishing degree.

Pretty much the only sound from the Sig would be the mechanical noise of the shells ejecting and sizzling in the snow. She popped in a full 30-round magazine and chambered a round.

After learning about the mission, she knew she and her team needed a rock-solid reliable weapon that was deadly at close range. For the weapons-training portion of her team-development period, she had coached her men to become experts in close quarter combat with the MPX.

Tango thought about this mission. As a top-notch mercenary, publicity-phobic governmental agencies or wealthy individuals, who valued secrecy more than success, typically arranged her jobs. However, she had nearly always had a face-to-face with the person who required her talents. This time was different.

Years ago she had an Alaska job like this. The work had come through a rogue Homeland Security director. It had turned out disastrous because her commanding officer had let the mission turn personal. This would never happen with Tango. She never got personal.

Tango thought about her upbringing in a Romanian orphanage and adoption by a wealthy childless couple from Atlanta. Throughout her early American schooling, her teachers had told her parents that she lacked empathy due to her "unbonded" early years.

In high school, Tango had observed and learned to act like she cared about relationships. As a skilled mercenary, her lack of feelings for other people enabled her to do her job more effectively.

As she installed the side-folding stock to her weapon, she ticked off the successful missions that allowed her to command the highest mercenary pay of anyone else in the business. She had recently commanded deadly jobs in the Baltics, South America and Africa.

The last Africa job had required her to kill everyone at a meeting of some well-protected arms dealers who supplied technology to middle-east radicals. Tango was pretty sure the French government had paid for that one, but she didn't really care who paid her.

A person who used voice-altering technology to sound like a slow-talking Johnny Cash had set up this current mission over the phone and through encrypted email. He had also supplied her with a team of raw soldiers who were so committed to the mission, she was sure they'd commit suicide if so ordered. It made her a bit uneasy working with them. They needed a much lighter touch to motivate.

Just then her secure satellite phone buzzed. Since starting this job, the phone was the main conduit for communications between her and her boss.

"Tango, can we talk?" Johnny Cash's slow deep voice came through the connection.

"Yes, Sir. I'm alone. Everything is going as planned. I emailed the report last night."

"I read it. Good job."

"Sir, I wanted you to know that some more civilians are scheduled to arrive here. Dylan Baker is one name on the list."

Tango thought she heard the boss take a slow but sharp intake of breath. "Your orders were to avoid taking life as long as things were proceeding according to plan. I just wanted to know if anything's changed due to the presence of people other than the university teams."

"We talked about this. The university people are not to be considered civilians even though they are non-combatants. They can cause more harm to our objectives than any other weapon."

"Yes Sir, and what about the students and interns?"

Tango attached a carbon-fiber handrail to her MPX while she held the phone with her shoulder.

"Sometimes we have collateral damage. It can't be helped," said the Johnny Cash voice.

16

AFTER AN HOUR, the steady Beechcraft Bonanza V-tail lulled Dylan to a peaceful light sleep. As he leaned back dozing, he wondered if he should engage Flapjack in a conversation just to keep the pilot alert.

"So, how long have you been flying?" Dylan ran his eyes over the mint-green retro look of the plane's interior controls and gages.

"I've been flying for 40 years. Started flying professionally in Switzerland for a courier service. Did some stunt flying and biplane racing then fell in love with an Alaskan girl. What about you? Are you a real native Alaskan or did you marry into it like me?"

Dylan adjusted the large noise-canceling headphones he and Flapjack were wearing. "I'm married to a pretty Hawaiian-turned-Alaskan girl. She's a journalist and working on an Africa story for a glossy magazine."

"So you don't see her very often?"

Dylan didn't see Suzie often enough. Ever since her writing career took off, she'd been on the go, and her travels had become a source of contention for the couple. Dylan wondered if living in a log cabin in the Alaskan woods was just too small a place for her. Her current assignment only allowed her spotty opportunities to call or email Dylan. "I'd like to see her more, but she's advancing rapidly in her writing assignments, so she needs to travel."

"Why don't you get her to write books? She can do that anywhere." Flapjack scowled at the clouds in the distance as he talked.

Just then the plane lurched. It didn't upset Dylan, but Flapjack's scowl deepened.

"Those damn thunderheads are throwing a gust front all the way over here." The V-tailed Bonanza lurched again.

"Are the gust fronts dangerous?" Dylan examined the clouds ahead. It looked like a string of pretty popcorn balls arranged on the horizon as far as he could see.

"Damn straight they can be dangerous, but nothing compared to what you see in front of you. The vertical wind shears inside a cell where up-drafts and down-drafts are side-by-side, can easily tear a huge aircraft apart.

"Sudden severe downdrafts can slam a 747 into the ground. And these suckers can be big. A downburst may be larger than two miles in diameter."

"Then why are we flying toward them?" Dylan thought about their plane's reputation for coming apart when pushed past its limits.

"I've never had a crash because I'm a cautious flyer. Look at those two cells that appear to be about 40 miles apart. I can fly this baby between them. It's going to be bumpy, but this plane can do it. Tighten your seat belt."

As the plane rumbled on, the bouncing and lurching increased and so did Dylan's discomfort. When dangling from a sheer rock wall and cycling his breathing, Dylan felt relaxed and in control. But he always felt a bit uneasy and out of control in a small plane.

"We can always go back. I had all three tanks filled up," Flapjack assured Dylan as he pointed the plane toward a big gap in the line of thunderstorms.

"What about lightning?" Dylan asked as he observed flashes within the thunderheads.

"I'm not too worried about it. You can never tell what it will do, but it's far more likely to fry our electronics than hurt us." Dylan did not find this comforting as he watched another amazing display of lightning bursts among the storm cells.

Flapjack reached around his seat and handed Dylan a bag with some saltine crackers in it. "The crackers will settle your stomach and you can barf in the bag. Don't barf on my new cockpit."

The plane was bucking and lurching and the tail seemed to wag as it approached the clear section of sky between two huge thunderheads.

Suddenly Dylan could hear the shrill stall horn announce that the plane was losing lift. The altimeter showed they were descending.

"Holy Shit!" Flapjack screamed. "There's a bigger cell hiding behind these. Let's get out of here."

As the stall horn became a scream, Dylan noticed that the Beechcraft's controls seemed to be having no effect on its performance. "What's going on?" Dylan yelled over the noise.

"We're in a downburst. I would have never thought it possible this far from a storm. These storms get stronger every year." Suddenly the plane lurched like someone was shaking it. The satellite phone came out of Dylan's lap and hit him above his eye. Crackers scattered around the cabin.

"We got to get out of this air." Flapjack pushed forward on the control arm, but the plane went nose up instead of nose down. Dylan felt himself weightless and wished he had eaten more crackers.

"Crap! We're going into a spin." Flapjack glared at Dylan, "Are you sure you aren't a doctor?"

Dylan felt the nose of the plane go down and could see deep forests and mountains spinning around. He wondered if the tail of the plane would detach. The altimeter spun crazily as they lost altitude due to the downward moving air and the nose-down attitude of the plane.

He saw Flapjack add power. "Hey, aren't we already past the speed tolerances for this plane?" Dylan yelled.

"We need control. The plane is falling about as fast as the air around it. We need to go faster than the air. I must get control." Flapjack didn't want to tell Dylan that if they suddenly hit still air, it would be like crashing into a stone wall.

Dylan heard Flapjack talking to himself. "Nose down. Opposite rudder. Avoid yaw. Slow climb. Slower climb." Dylan watched Flapjack put down the landing gear. No place to land, but he guessed the old pilot now wanted to slow the plane without using the flaps.

Just as the plane seemed ready to smash into a forest, it steadied and started on a slow climb. Flapjack put the nose of the plane into a very slight climb. "Got to get this speed off slowly or we'll rip apart my Beecher." Just above the treetops, the plane made a gradual turn away from the storm.

"Holy shit! I think we almost died," said Dylan, a note of near panic in his voice.

"You're probably right. We're not going to be able to land at Lake Minchumina Airfield like we were told.

I'm heading far southwest to get away from these clouds. We'll land at Chuathbaluk Airport. It's a damn gravel runway, but a family I know can put us up. Do you like dogs?"

"I love dogs." Dylan said through his mouthful of crackers.

The farther they traveled from the storms, the smoother their flight became. When Dylan saw Flapjack take the plane into a landing pattern over the waters around Chuathbaluk Airport, he felt ready to kiss the ground.

Tomorrow he'd have to get back into the plane and fly to Lake Minchumina. Dylan wondered why a tiny village like Minchumina would be so important to the governor and Anja.

17

THE UV FILM-COATED WINDOWS of Governor Hardy's Anchorage Mansion reflected light like a mirror. During the day much of the bright sunlight was directed away from the offices. After dark the interior office lights reflected back from the windows creating a large reflective surface.

Harold Cheek paused in front of his reflection and glanced at the coffee stain-dotted dress shirt peeking out from his sport coat and his slightly disheveled black hair. As Senior Political Advisor, Cheek wanted to portray an independent and creative image, but some people just thought he looked sloppy.

Cheek felt he perfectly reflected the image of a savvy campaign manager fresh off a big victory in Louisiana where he got on-line charter schools locked into state law despite their terrible educational outcomes. The fact that the Bates-Hardy campaign had agreed to nearly all his terms filled him with confidence.

Cheek again glanced at his reflection in the tinted windows and sucked in his bulging stomach. Satisfied with his suddenly lean physique and 5'11" height, Cheek looked down at his desk for any urgent messages that might have appeared during the budget meeting. *Nothing.* He was very pleased with the results of today's meetings.

Yes, it appeared that he had conceded to the committee's proposals and given up some measure of economic flexibility. But the committee probably didn't realize Cheek had gained strategic control at a time candidate Greg Yeung's campaign for governor was ramping up and committee control could make a huge difference.

A glance at his watch indicated twelve minutes until his meeting with the governor. The stack of papers dedicated to the next meeting contained sticky notes and tape flags identifying key points. After a quick review, Cheek took a moment to consider the situation. The University of Alaska's Environmental Forum was going to be a focal point for the Governor's re-election efforts and today's meeting with Governor Hardy would focus on both strategy and details.

The self-assured Colonel Bolton had nominated himself to represent the oil industry. On the up side, whoever represented the Governor during the forum debate with Bolton had the chance to connect the environmental and commercial dots for the public. The huge potential risk was that a poorly worded statement or rebuttal would fan Yeung's oil subsidy programs. The governor's representative would need to be charismatic and articulate.

IN ANOTHER PART OF THE MANSION, Governor Leola Bates-Hardy checked her morning calendar noting that her political advisor should appear within a couple of minutes. Cheek was well qualified and had the political acumen she needed to out maneuver Greg Yeung's high-priced team of experts. Cheek's strategic input had historically proved to be right on the mark.

An odor somewhere between old gym socks and a garden compost bin preceded Cheek as he entered her office.

"Good Morning, Harold." Governor Bates-Hardy smiled at Cheek's punctual arrival as she directed him to a chair somewhat distant from her desk as he often brought in a strong odor when he entered a room.

"Good morning, Governor. Thanks for taking a few minutes to meet with me." Cheek ignored her offer and stood in front of the governor's desk.

He leaned over her as he spoke, "I've been thinking about the Environmental Forum debate. Have you decided who will represent our side in the keynote session with Bolton?"

Bates-Hardy had been struggling with that very question. "Well, Deputy Assistant Secretary Peterson has had the training and comes across very well."

"I thought that might be your line of reasoning. I would like to represent us in the debate."

"Harold, are you nuts? This isn't just a critical juncture on the campaign trail; it's the only annual environmental forum. Does this debate warrant your participation? If we bring in our biggest guns, won't your presence give the impression that Yeung is operating from a stronger position?"

"Yes, it does." Cheek went on to explain his strategy. Within a few moments, Leola realized that Cheek had hit the mark again. The man was brilliant.

18

THE ZETROS COMMAND CENTER appeared to be a little crowded, but Tango and the two technicians were just going to have to make due.

The grey static seemed to clear as the listening devices came on-line. The team's operative was well placed in the ARCS security team and had complete freedom to move throughout ARCS. This transmitter was optimally placed within Professor White's office, placement worthy of a commendation in her next report.

The limited space of the Zetros and the muted voices coming through the speakers forced Tango and her team to line up shoulder to shoulder to catch each word.

An authoritative, precise voice filled the Zetros. *Professor Dave White*, Tango thought.

Dave's voice boomed, "I agree. The results are within two standard deviations of the expected results. That's within the acceptability limits. Soon the Scrubber will be operating efficiently, and we can start saving the world. " Papers shuffled and a chair was repositioned.

Tango shook her head. "Something seems a little off. Two standard deviations is not very impressive, especially when it is relative to an 'expected result'. Who knows whether the expected result is an appropriate research target? Somebody could have arbitrarily pulled the 'expected result' out of his or her butt."

"Not more data is needed to think, Professor?" The speakers boomed out Zhang's words.

The authoritative voice responded, "More data is always better. But, we are under the pressure often felt by applied science, the pressure of a deadline. Why don't you go back to the lab, review your data and recheck your calculations? Maybe there is a simple experiment we can quickly run that will provide us with additional information. Thanks everyone. Let's meet again tomorrow morning."

Chairs scraped the floor and papers shuffled. The mumbling and soft conversations faded with the closing door.

Tango and the technicians leaned back from the huddled formation that surrounded the speakers. "OK, I'm going to take a break." Tango nodded to one of the technicians, "Take the next shift. Give a yell if any relevant conversations start up. And double-check the recording equipment. The Boss will not want to miss the smallest grunt." The technician nodded, realigned his oval, off-centered glasses, and settled into the chair.

Tango lingered for a moment. She had been on too many missions where some team members were unclear as to the objectives. It was essential that she stay in close contact with the technicians and then relay information to her team. Updated, unedited information was critical to optimal team performance—especially with her current soldiers.

Mentally Tango reviewed The Boss' instructions. *The University people are to be considered hostile non-combatants rather than civilians. Avoid taking a life, but do not sacrifice the mission objectives. Collateral damage may be unavoidable.*

FIVE HUNDRED YARDS AWAY, Professor Dave White set his heavy-rimmed glasses on his desk and ran his fingers over his polished bald scalp. He was a personable, scholarly type with a poorly hidden receding hairline that gave him the appearance of a high school science teacher. His pear-shaped frame and ever-present lab coat hid the fact that he was an expert handball player.

Dave glanced around his office noting that he really didn't mind his Spartan surroundings. It was pretty much a desk, a chair, and a couple of family pictures on the wall. But at the ARCS, with office space at a premium, Dave was grateful for a quiet place to think.

Zhang was right. The data was insubstantial. But the variability in the data wasn't the only thing bothering Dave. The remaining time and financial resources were at critical levels. He had to figure out a way to efficiently collect the data needed to verify performance.

Replacing his glasses, Dave exited his office and headed down the hallway to the break room. The knotty pine-framed windows sometimes helped him forget that he spent nearly all of each day in his labs.

The windows provided a view of the hillside and forests. The snow outside contrasted sharply with the rugged hills leading up to the higher elevations and clouds that perpetually obscured the views of Denali's summit.

Dave's internal conflict bubbled up like the hot water on the stove in the break room. As he prepared his herbal tea, his mind once again ran through options for accelerating the research. *Perhaps I should create a spreadsheet that prioritized the performance issues based mostly on technical significance, resources required and time to completion. Maybe we will just need to pick our battles.*

But for the next few moments he must set research prioritization issues aside and focus on the matter at hand, the upcoming visit by Dylan Baker. As a graduate student, Dylan had stumbled on amazing soil bacteria that could change carbon dioxide into a flammable alcohol.

Dylan's brief stint in the lab as a graduate student had some interesting consequences. His lab experiments hinted at a methodology for converting exhaust carbon dioxide to a useable alcohol. By bubbling exhaust through a bacterial solution in a glass tube, he was able to show lower CO_2 levels in the exhaust and confirm the presence of combustible alcohol.

The implications for the greenhouse-gas-generating coal industry were staggering. Dylan's initial work provided the biological foundation for the performance studies currently in progress at the Alaska Remote Climate Station. Dave's team created a healthy growth environment for the bacteria and collected the useful by-products.

The break room clock chimed the hour provoking Dave to take a last sip of tea. Dylan Baker was arriving.

19

PROFESSOR DAVE WHITE SCURRIED down the hall nodding to colleagues and deliberately avoiding a glance at the banners lining the walkways, "Welcome Dylan Baker" and "The New Coal Industry Begins Today" and "Coal is Turning Green". Balloons fashioned from plastic gloves and strips of colored paper adorned the hallways.

And there was indeed cause for celebration. The consequences of Dylan Baker's original work may start a better future for both the coal industry and the earth's environment.

Dave rounded a hallway corner and passed under the 30 foot-long L-shaped, foam-filled cylinder suspended from the ceiling. The multi-layered matrix formed the technological basis for ARCS' project to remove CO_2 from coal plant emissions

"Good morning, Dave." The tan lab coat worn by university volunteer interns barely hid the intense enthusiasm of the young woman turning the hallway corner. Tall, with light brown hair that matched her lab coat, Lindsey West was regarded as a brilliant, upward bound scientist whose vivid presence brightened any dour lab.

Although Dave had provided the technical trigger for solving some of the project's design issues, Lindsey's biochemistry insights had shaved years off of the project.

Besides the alcohol removal, Dave's contributions focused on *the Scrubber*'s physical design and the flow of nutrients and wastes. Additionally, Lindsey's insight helped fine tune the nutrients and maximize growth.

"Good morning, Lindsey. I glanced through your report. Recheck Table 3, but otherwise, very nice."

Lindsey beamed her pretty smile as she turned and headed down a hallway.

Dave rubbed his shiny bald head and turned his thoughts back to Dylan Baker, a man of contrasts. On one hand, Dylan had provided preliminary discoveries for a device that could place a green environmental stamp on the coal industry's toxic carbon dioxide emissions and provide a cheap fuel additive.

On the other, Dylan's daughter, Anja, was the lead fundraiser for the environmental organization, Mountain Club. Dave had heard that Anja Hart was pleasant to look at, intense, incredibly frank and effective in her use of the Mountain Club's funds.

Dylan must be walking an interesting network of tight ropes with Anja and the forces of environmental protection on one side and the advocates for coal development on the other.

And where would woodsman Dylan fall, on the side of abundant energy with global pollution or higher energy costs and environmental consciousness?

Perhaps Dylan thought his discovery would put two typically opposed groups on the same side.

20

AFTER AN UNEVENTFUL EVENING FLIGHT with Flapjack, Dylan paused at the ARCS facility entrance his duffle at his feet. He took a few minutes to collect his thoughts and recover from the stress of the journey. The Governor's instructions reverberated in his head: *observe the testing, review the results and summarize your comments in a report to her and Anja.* That should be simple enough: *if I had followed the research over the past 15 years.*

Instinctively, Dylan began oval breathing, his calming technique that helped with mental focus during times of physical or mental stress. Inhale through the nose, pause, exhale through the mouth.

The main door opened to reveal a middle-aged lab-coated scientist. "Dylan Baker, it is a pleasure to meet you! I'm Dave White." Dave's handshake emphasized his enthusiasm "Thanks for coming. It is a real honor to have you here." As they started walking down a hallway, Dave said, "I hope the storm wasn't too bad."

"Well Dr. White, we made it." Dylan looked down and brushed the cracker bits and dog hairs that continued to adorn his shirt.

"Please call me 'Dave'. Yeah, I'm sorry about the scheduling. Research-wise we are at a critical juncture. The team will find your presence and support to be invaluable. Add the pressure from the election and a series of weather fronts agitating our happy little get-away, it seemed best to shuttle you here as quickly as possible. Thanks for acquiescing to our timeline. We'll have something for you to drink at the conference room.

"Dylan, you are going to have to accept the fact that you are a bit of a celebrity around here. As you know, your initial work contributed to the scientific foundation for *the Scrubber.* "

"You know, I'm not totally clear on how you and your team made the jump from my initial ideas to a functioning prototype. I'm really looking forward to hearing the back-story." For the first time, Dylan noticed Dave's shiny head and wondered if he polished it.

As they rounded the corner, Dylan rolled his eyes as he saw the *Welcome Dylan Baker* sign. A few steps further down the hall revealed several pro-coal banners. Dylan began to connect the dots in Governor Hardy's strategy.

Dave walked ahead. "First we pushed back bedtimes so our science team could meet you. Then you and I will have some time to answer your questions." Dave led the way to a conference room, where five scientists looked up curiously at Dave and Dylan. "Everybody, I'd like you to meet Dylan Baker."

Seeing a room full of new faces, Dylan had a moment of panic. *How can I possibly remember all their names?*

"Dylan, this is Sully Meyers, our geologist and part of our mechanical engineering team." Dylan smiled and nodded towards the chubby, disheveled, stereotypical image of a distracted scientist. Sully's white-framed glasses contrasted with his dark African-American skin and shaved head.

Dylan could not remember ever seeing white-framed glasses before.

Sully, with lab coat buttons in the wrong holes and a buttery Pop Tart in hand, acknowledged Dylan with a smile and friendly nod. He seemed to be preoccupied by the butter running down his wrist and had somehow had put a greasy fingerprint on one lens of his glasses.

"And Dylan, this is Ethan Rivers." With his slender build, dark curly hair, heavy eyebrows and big brown eyes set off with light mascara, he reminded Dylan of the eyes of a stranded seal pup.

Dave continued, "Ethan is the other half of the project's mechanical engineering team. He's just finishing up his post-doctoral research."

With Dylan's acknowledgement, Dave moved on, "Next, I would like to introduce Zhang Qi, the project's biochemist." Zhang had long slicked-back hair that reminded Dylan of an Arctic fox's summer fur, a wide friendly face, and a broad smile. He greeted Dylan warmly.

"And also, I would like to introduce Taylor Phillips. She's our microbiologist." Taylor looked athletic, mid to late 30s, and her attractive face perfectly framed by her long dark brown hair and eyes that suggested some Asian ancestry.

To Dylan, Taylor looked like a serious and competent scientist with just a touch of controlled sensuality. Her smile seemed warm and genuine to Dylan. As the only female on team, Dylan thought he could remember her androgynous name. He wasn't sure he could keep track of Sully-white-glasses or seal-pup-eyes Ethan.

Shuffling over from a refreshment area was a very short, thickly built man putting down a coffee cup with hooked metal grippers on the end of a prosthetic arm. The whole right side of the man's body appeared to be misshapen except his friendly, boyish face. "So you're Dylan, the-carbon-scrubber-man. I'm Conrad, the computer man." He extended his prosthetic hook for Dylan to shake.

Not quite knowing what to do with the offered prosthetic hook, Dylan reached out to shake the metal pincer.

Quickly Conrad withdrew the hook and took Dylan's right hand with a strong left hand. "We are so impressed with your initial research. Funny that someone hadn't taken it and run with it before now."

"Thanks, Conrad the computer man," Dylan said.

With a look towards the conference room clock Dave said, "Well it's getting a bit late, why don't we adjourn until tomorrow morning. And Dylan, I'm sure you could use a break." Dave noticed a consensus of affirmative responses. "Taylor, would you show Dylan to his quarters?"

Taylor looked at Dylan and smiled. "Very good, I'll see everyone in the lab for tomorrow morning's test run." Dave turned into the hallway followed closely by Sully, Ethan and Zhang.

Taylor and Dylan headed in the opposite direction. Taylor smiled again, "You know Dylan, I have been really looking forward to meeting you. Let me ask you a question first. How does a scientist become an alpine guide?"

"Well, initially I believed that science was my passion. But in graduate school, I realized that research science involved way too much time in an indoor lab. Every summer I climbed mountains and guided trips and made good money. It was pretty easy for me to see which path was better for me."

Taylor smiled and nodded. "I ask because you live the life I believe I might have selected for myself. You are an alpinist and a woodsman. And on top of that, you are a closet scientist who has the cognitive resources to understand technical problems."

"I think you are overestimating my cognitive abilities," said Dylan transferring his duffel to his other hand

"I've never had the courage to explore the wilder side of my personality; you know like you've done . . . away from towns, cities and civilization." Taylor paused, a little embarrassed. "I'm sorry. A little too much information?"

"No, it's fine. Tell me about your science background."

"Well, as Dr. White indicated, I'm the local microbiologist. As a lover of tiny bugs, I want to know how you just dug up some dirt and found a bacteria like *Ralstonia*."

Dylan started to speak, but Taylor's enthusiasm kept him from forming his response. "I also love science and field work. And, in the ARCS resource-limited world, I'm also the local EMT—available for any medical emergencies that may occur."

"I have so many questions to ask, but I'm really beat. When does the staff eat breakfast? Could we meet tomorrow morning?"

Taylor smiled, "Sure. Generally we begin rolling in around seven."

Dylan decided he wanted to see her smile again.

21

BACK IN HIS OFFICE, Professor Dave White settled in his chair and collected his thoughts. Dylan's arrival and introduction to the technical team seemed to go well. This ARCS project was in fact the practical application of Dylan Baker's initial CO_2 scrubber research.

Dave glanced at his biochemical vision hanging, framed, on the wall:

Carbon Dioxide $\Rightarrow$ **Branched Alcohols** $\Rightarrow$ **Gasoline Additive**

First, Dave's team had identified *Ralstonia eutrophia* as the bacteria that converted a mixture of carbon dioxide and hydrogen gas to alcohols in Dylan's early soil samples. The team then purchased a genetically modified *Ralstonia eutrophia*, known for improved scrubbing and alcohol production, from researchers at the Massachusetts Institute of Technology.

With the development of a special filter, *Ralstonia eutrophia* found a home where it could actively convert endless amounts of CO_2 to several different alcohols. These alcohols, destined for use as gasoline additives, were elegantly funneled away from the growing bacteria.

But the progress towards Dave's vision had slowed. Although his team had been showing a good front, Dave knew that the pressure and the timelines had been taking their toll.

I hope Dylan's presence energizes the team, Dave thought.

Dave's tablet was displaying a collection of news media reports. The headlines caught his eye and produced a grin. For once it seemed like Dave was getting a break.

News accounts indicated there was considerable federal legislative energy focused on carbon dioxide emissions produced by coal-fired power plants. Federal agencies were vetting coal emission guidelines for new power-generating coal plants and planning to install new regulations to limit pollution from existing coal plants.

With a little scientific luck, a successful test run tomorrow, and a creative analysis of the data, Dave might just be able to deliver on the Governor's assignment. Then after the election, with more time and resources available, he would be able to enhance and document *the Scrubber's* performance.

Dave noticed a moldy tea bag sitting in a spoon behind his computer monitor. As he dumped the bag into the trash, he imagined how pleased the governor would be with his efforts.

22

FROM HIS VANTAGE POINT on the main stage, Colonel William Bolton surveyed the audience returning to their seats at the end of the break. The cameras dotting the elevated stages in back of the auditorium were for the most part unattended. There had been hope that the Governor Hardy's environmental plank would add some media attention to this year's University of Alaska Environmental Forum. Certainly no attendance records were being broken this year.

Bolton scanned the stage. The forum's moderator, a heavy blond woman, Beverly something, was reviewing her notes from the moderator's table. From the subtle curves accented by her sweater, her last name certainly wasn't Hills.

Bolton wondered who the governor would send to this important debate. It would probably be a low-level staffer with a college class in debating. *These Alaskans are such hicks.*

On the opposite side of the stage, a tall, angular figure ascended the steps. Bolton's speculation stopped dead, mid-thought. *That was Senior Political Advisor Harold Cheek! What a message! The oil industry marketing efforts must be hitting a nerve, because the Governor is obviously on the defensive. Why else send her* Senior Political Advisor? Bolton couldn't imagine a better outcome. As the audience recognized the governor's SPA, cameras and reporters seemed to appear out of nowhere. Cheek's presence immediately doubled the media attention.

As Cheek settled in, Beverly called the University of Alaska Environmental Forum to order. After the speaker introductions and an overview of the forum's format, she asked Bolton to begin with his prepared remarks.

Bolton mentally confirmed his erect posture as he projected his successful military and business presence. "Mr. Cheek, as a senior political advisor, you are very well aware of the oil industry's benefits to the people and State of Alaska. This industry has created cheap energy, plentiful jobs and the revenues to develop infrastructure, and regular payments to financially protect the citizens of Alaska."

Harold Cheek seemed a little distracted. Beverly glanced at her notes and then Cheek. The audience waited patiently for Cheek's response.

"Yes, Colonel Bolton. The petroleum industry's contributions to the State of Alaska have been profound." Cheek paused for dramatic effect. "But you cannot deny that those contributions have come at a dear price. I'm talking about the environmental cost."

Bolton smiled inwardly. "This cost is gladly covered by the petroleum industry. Whenever some problem has come up, billions are spent in fixing the situation. But there is a trade-off, a small portion of nature and a little environmental damage in exchange for the benefits of a modern society. Since humans cut down the first tree for firewood, we've been willing to bear the cost of keeping warm versus leaving the tree God put there for us."

Cheek leaned forward. "But, Colonel Bolton, you are minimizing the twin environmental impacts of the pipeline itself. According to the latest Pipeline & Hazardous Materials Safety Administration, on average, U.S. pipelines spilled over 3.1 million gallons a year between 2008 and 2012. Add to this the disruption of migration patterns and the local wildlife."

Bolton nodded in acknowledgement, "Well, the petroleum industry is actively responding to both of those environmental impacts. Field evaluations of enhanced sensors that respond instantly to minute pressure changes are very encouraging. The petroleum industry has incorporated accurate satellite migration data into environmental impact analysis data. Responding to our commitment to the planet we call home, the oil industry has become radically proactive regarding these issues."

"These activities seem rather cosmetic. What if the price of oil drops dramatically like it did not too long ago? Alaskans had to pay income tax instead of getting paid by the state government. They didn't like that!"

"Very unlikely." Bolton shook his head. "I don't see it happening again. The world petroleum producers have been down that road before. This time they will be ready if a price drop occurs again. We'll adjust production to maintain the price of a barrel of oil. Saudi Arabia and Iran have allowed lowered production to support prices. And even if nothing is done, the price of oil always bounces back."

"Perhaps we should consider an alternate energy source like . . . coal."

Yes! Bolton felt the thrill of a chase leading to the kill.

"Well, I wasn't going to mention coal. Mr. Cheek, but since you did bring it up I do have a couple of comments. As we all know the environmental impact of power generating coal plants is truly extraordinary. To start with you have the health and environmental impact of nitrogen oxides, particulate matter and mercury.

"All three of those pollutants have been linked to health, tissue and lung issues. Coal burning energy plants also emit cadmium, carbon monoxide and arsenic. But those are just the tip of the iceberg. The big players are sulfur dioxide, which is responsible for acid rain and damage to crops, lakes and streams. And then there is carbon dioxide which is the primary cause of global warming."

"What if there was some new technology that could effectively remove carbon dioxide from coal utility emissions?" Cheek picked at a cuticle.

"That's absurd. Coal technology research has been around for nearly as long as coal has been an energy source. There is nothing like what you describe. You're talking a vision—or more like an hallucination." Bolton chuckled to emphasize his point.

"But what if there was some new innovation?"

"There won't be any new innovation. Petroleum and natural gas are cheap and less polluting than coal. Anyone waiting for some magic to make coal clean is delusional." Bolton filled the room with his confidence.

Cheek smiled a huge victory grin for the first time. "OK, Colonel Bolton."

GOVERNOR LEOLA BATES-HARDY looked up from her tablet, where she'd been watching the forum, and gazed out her office windows. *High marks, Harold. You played Bolton like a puppet-master.*

23

THE NEXT MORNING TAYLOR and Dylan walked through the bustling ARCS hallways, through the outside door, across the snow-covered patio to the door labeled Lab 4.

"This is our component lab. It contains a scaled down working model of the coal plant exhaust stack. Exhaust gases are injected in one end and then are monitored as they exit the other end of the smokestack, a practical design to test the effects of a device that may significantly improve air quality on earth."

Taylor handed Dylan a tablet, which displayed several spreadsheets. "Now before you get all excited, remember that these data are from our earlier runs. We were still fine-tuning the systems. Put on your 'bigger picture' lab glasses."

Dylan found himself impressed with Taylor's ability to project an approachable yet professional image. Realizing Taylor had stopped talking and was waiting for comment, he mentally replayed her words and glanced at the spreadsheets and the massive equipment that filled the lab.

As if sensing his warm thoughts, Taylor smiled and led Dylan over to a tubular structure shaped like the letter "L" lying on its side.

"I've seen this design. Governor Hardy has a model of this equipment in a box on her desk."

"Well Mr. VIP, next time get me a cell phone picture," Taylor chided.

The 12-inch diameter, 20 foot-long horizontal part of the foam-filled tube was studded with input gas sensors and regulators.

"Exhaust gases are generated at the remote combustion station about a quarter mile from here, compressed and stored in tanks. Those tanks act as the source of the gases we heat and run through the LCM."

"LCM? Dave mentioned that yesterday. That's the filter part of *the Scrubber*, right?"

"Yep. Layer Cake Membrane—that's our pet name for the mesh, inner filter that allows the bacteria to work their magic. The bacteria embedded in the membrane filter use carbon dioxide and hydrogen gas for energy and growth. A good analogy might be that membrane filter is the engine that will power the carbon-scrubber."

Dylan considered all the acronyms and names as he studied the 90-degree elbow characteristic of coal plant exhaust stacks and the 30-foot vertical section of the foam-filled tube.

Following Dylan's gaze, Taylor's commentary continued, "The sensors at the exit ports located at the top of the vertical section monitor the "scrubbed" exit gases before they are then captured and stored. Exit gases from the earlier trials contained very high levels of pollutants—a bit of a disposal challenge. I believe ARCS contracted with an outside company that slowly vents the gases in downtown Beijing where they could blend in." Dylan's puzzled look was quickly defused by Taylor's teasing expression and ear-to-ear grin.

Taylor shepherded Dylan outside and back across the patio to the Lab 4 Control Room. Inside, Dylan confronted a massive wall of monitors, gauges and cameras.

"What do you think?" beamed Taylor.

"Impressive!" Dylan tried to take in the large amount of information presented on the wall. "I have a couple of small questions: How do you know that the "L" provides an accurate representation of what happens in a real coal plant exhaust stack? And how did you solve the alcohol removal problem?"

"You think we just sit around here making this shit up?" Taylor's eyes signaled the smile she tried to conceal. "Yes, all of our performance data indicates that the "L" provides a good working model for what happens in an exhaust stack. But things haven't always smelled like roses.

"For example, we learned that we needed to preheat the injection exhaust gases and then control the gas temperatures throughout the "L" to more accurately mimic a real exhaust stack—temperature dramatically influences microbial growth. We are beyond testing the "L" and are now focusing on *the Scrubber's* performance.

"Dr. White gets the 'alcohol removal award'. His partial-vacuum osmotic filtration system removes nearly all of the alcohol produced by the bacteria without harming the culture." Taylor's pride at being an active member of the team was evident.

"As far as the results go, we frequently see a significant reduction in carbon dioxide levels as the exhaust gases pass through *the Scrubber*. But, on occasion, we have exceeded our exit gas CO_2 targets. Here are some of the results." Taylor handed Dylan her tablet.

Dylan quickly glanced over the spreadsheets then paused to review the conclusions. *These aren't bad results.* "So, what's happening today?"

"Today is our first full system test after two weeks of equipment calibration. Dr. White should be here any moment. You will be able to see *the Scrubber*'s performance first-hand."

PROFESSOR DAVE WHITE downed the remaining tea and took a deep breath. He'd been reviewing the test protocol since early today and everything seemed ready to go.

This morning the ARCS hallway was bustling with technicians and students. Dave walked through the doors of the Lab 4 Control Room to see *the Scrubber* and his development team present and accounted for. Off to the side, Dylan stood next to Taylor and tried to remember everyone's name. The team seemed focused on a technical issue. Dylan looked a bit lost.

Sully, a grease smear still on his white-framed glasses and Zhang were huddled over *the Scrubber* and its sensors in the main lab. The control room monitors indicated that the hydration system was engaged and maintaining the proper levels of moisture for the membrane's growing bacteria.

The carbon dioxide and hydrogen gas feeds were ready to inject gases into the cylinder. The CO_2 sensors at the opposite end of the "L" were capable of measuring the smallest changes in the gas concentration.

Ethan-engineer-seal pup had moved over by *the Scrubber* module and started fidgeting with a turbulence damper. Taylor hovered over the control panels. Dylan stood by and watched, awkwardly trying to stay out of the way.

With an acknowledgement from both Sully and Zhang, Dave decided everything was ready for the test. "OK, Ethan, hit it!" Dave's eyes bounced from screen to screen.

Ethan pressed several buttons on the large console. "Hydration system initiated. Twenty minutes until fully deployed. 'Come on my little pretties.'"

Taylor responded to Dylan's confused expression, "The hydration system keeps the bacteria alive and happy in the warm, dry smoke stack environment." Dylan nodded his appreciation.

The scanners followed the progress of the wave of moisture traveling up the LCM. "Hydration is complete. I'm opening up the gas jets." Excitement crept into Ethan's voice.

Everyone studied the monitors.

"Yes!" Zhang exclaimed. "Look. The quick bacteria dropped CO_2. Today better. Yesterday not. Hm, why quick bacteria stopped? Ethan, your pretties are not so. CO_2 jumping up and down at the exit sensors."

Dave White exhaled some of his apprehension. He mentally completed the calculations. *Although CO_2 levels initially dropped at the exit sensors, they began to rise. These results weren't that bad—certainly sufficient to warrant additional test runs.*

Frustration overcame Dave's typical self-control. "What is the problem here? You guys are some of the most capable people in your field!" Dave's voice crescendoed with each sentence. "This experiment is progressing like some high school science project where any pitiful progress is acceptable. Come on team, figure it out!" Each word was succinctly articulated. "Do I need to remind you how tight the timeline is?"

Dave quickly exited the lab. The team stood frozen in place at this unexpected outburst. The last time Dave exploded at a research team, heads and careers rolled. A wordless shared commitment passed silently between team members. Everyone turned to their notes to recheck calculations and assumptions.

Zhang was first to speak to the team, "I today parameters build. I open with your thoughts."

DAVE'S STEPS SOUNDED a little heavier as he headed back to his office. *I shouldn't have blown up at them.* He tried to calm himself but they were so close and there was so little time. The experiment's results reverberated throughout his thoughts.

In his mind, Dave ticked off the three reasons why this experiment should be working better: The biochemistry was sound and well documented. These bacteria naturally grew well in the soil and consumed CO_2. And computer models suggested *the Scrubber*'s prototype should be able to first reduce CO_2 levels and then detect changes in CO_2 concentrations. *It should be working better. It has to work better!*

Back at his desk, the tea was steeping as Dave reached for the lab notebook on his desk. Dave took a deep breath, opened the notebook and began reviewing the pages.

Dave sipped his tea and examined the notebook's pages. He had to be careful. The formal review by the Science Grant Committee was already on the calendar. The pressure was on. *If only we had a few more weeks before the review,* Dave thought.

Dave shook his head. The data wasn't what he wanted to see. He needed more time to make this work. With some breathing room, Dave and the team could tweak *the Scrubber*, generate a large amount of conclusive test results and present all of the data. His little team could save the world.

It would be so tragic if the entire project were cancelled because a small portion of the data fell outside the required range. So tragic! All Dave needed was a little more time to confirm performance.

24

FLANKED BY TAYLOR, a puzzled Dylan walked through the ARCS hallways. Both Sully with his white-framed glasses and Zhang with his slicked-back hair followed closely behind.

Dylan's mind reeled. *Who would have thought that my meager experiment and preliminary theories would have evolved into this outcome? Scientists have been trying to measure and minimize or eradicate coal plant emissions for decades.*

Dylan remembered an article that stated a typical coal plant generated millions of tons of CO_2 per year. The implications were fascinating, the possibility of transforming coal power plants into facilities that were significantly more earth-friendly. And this transformation used lowly soil bacteria rather than an environmentally damaging device whose manufacture also left its own carbon footprint.

"Lump of coal for your thoughts?" queried Taylor.

Dylan grinned. "I am just amazed at how everything has come together. The biochemical pathways are really not that complicated. Dirt bacteria happily use carbon dioxide and hydrogen gas for growth, but I don't understand where the alcohol by-product came from. All I know is that, after a time, my bacteria didn't do very well."

Bubbling with enthusiasm, Zhang edged forward to be closer to Dylan. "You arrowed us correctly with bacteria in dirt paper. We buy *Ralstonia* for cooking our exhaust gas and became really interested."

Dylan nodded as Zhang continued, "Natural *Ralstonia* not efficient cook CO_2 to alcohol. From MIT, we get cooking to alcohol *Ralstonia* mutant."

"I overheard someone discussing 'branched alcohols'," Dylan said.

"Branched alcohols more complex than two-carbon alcohol in beer. Have taller energy almost like gasoine."

Dylan looked around at the team, "From what I can understand, you guys have done some really good work."

Each team member acknowledged Dylan's praise.

"There are still several major challenges," Taylor commented. "Maintaining living micro-organisms on the inside of the smokestack is no small undertaking. Hydration, growth factors, in-coming fresh media and removal of depleted media, and protection from the heat are all key concerns that need to be addressed. And, given all of the experimental variables, our ability to repeat the same experiment and consistently observe the same results tops our list of challenges."

Sully looked a little puzzled as he pointed a jelly sandwich at Dylan. "So Dylan, why are you here? Tired of woodsy living? You live in the woods just outside of Seward, right?"

"Yes, I live in a log home that's a short walk from town. But it is also nice to get out every now and then, especially into the world of science." *I wish I really did know why I was here.*

"I guess what I'm saying is that you dropped off the research journal circuit 30 years ago. Why show interest in your project now?"

"The governor has asked me to provide her with my perspective on the status of this project." *And maybe get Anja Hart's buy-in. I'm hoping I'll get a daughter out of this.*

Sully's brow furrowed as he took off his white-framed glasses and frowned at a grease mark on a lens. "OK, that makes sense. If Governor Bates-Hardy can find a way to gain support from the powerful coal industry lobby and not alienate the environmentalists, her re-election would be a much simpler process."

"Maybe you also pull governor's funding?" Zhang said as a long strand of black hair balanced on his ear.

"Yeah right, Zhang. The needed equipment for this project comes before the new investigations you envision!" Sully looked to Taylor for support. Taylor just shook her head in *don't get me involved.*

The group paused at the intersection of several hallways. Zhang grinned, "We tomorrow in the exhaust stack will create stable bacteria cooking. I know!" Zhang and Sully turned and headed down a hall.

Despite his exterior enthusiasm, Zhang could not help a feeling of gloom about the upcoming test.

25

THE EXTRAORDINARY NAH-GOON Tea Emporium was a favorite of both Bolton and the local students. Nah-goon, in Inuit means, *Which way*? was characterized as *local Alaskan culture meets the world of English tea* ambiance. Nestled around the corner from the University of Alaska campus, it was known for its expansive tea collection.

William Bolton acknowledged the hot tea with a nod as he finished his texting. Jane, a coffee person, had no drink on the table. She leaned forward to clear a space for her tablet. As Jane turned, her blouse opened slightly. Not only could Bolton see Jane wasn't wearing a bra, he thought he might be looking at her small breasts under a sheer camisole. Bolton smiled. He loved the cut of that blue blouse when paired with no bra.

Jane interpreted Bolton's smile as self-satisfaction as she squirmed in her seat across from her boss and waited patiently. The peaceful feelings she expected to be fostered by this mellow tearoom ebbed away as she thought about the Environmental Forum debate and what a dick Hardy's Senior Political Advisor was.

As Bolton was completing one text, another arrived. Bolton's distractions left Jane alone at the table.

Jane continued to analyze the source of her unsettled feelings. First, Harold Cheek, the governor's senior political advisor, appears opposite Bolton for the debate. The governor must have known that a high profile representative would signal concern or panic. And then Cheek's smile as Bolton skillfully summed the environmental negatives associated with coal-based energy production. Cheek even seemed to radiate success.

Bolton returned from his texting. "Please clear my evening schedule."

Jane checked the Super PAC calendar. "I'm not your personal secretary, I'm leading your Super PAC. Call Elizabeth and have her clear your schedule."

Unperturbed, and pointing to his phone, "This last text has changed my plans. Please tell my secretary to cancel the meeting."

Jane shrugged." That's not what you hired me for. What's up?"

Bolton puffed himself up, "I think our cash flow problems are over. I'm talking to some big donors tonight."

"We wouldn't have any cash flow problems if you hadn't approved ad buys without consulting me. You are spending money we don't have." Jane sat erect in her chair, still looking small.

"That will all be over with after tonight. These guys want Yeung elected, and I can make the case for campaign donations."

"And you don't want me there? Why did you hire me?"

"Now don't get all pissy. These people want to talk to me, not you. Some of our conversations are going to be just business, and you don't run my businesses, just the Super PAC."

Jane felt uneasy. She should be present at donor meetings. Bolton was back to micromanaging the Super PAC. *Why did he hire me if he's going to make the decisions?* Bolton's need for Jane seemed to be fading. Or was it his trust?

26

A KNOCK ON THE DOOR roused Dylan. "Ready to head over to the lab?" Taylor smiled at catching a glimpse of Dylan's repetitive movements.

"Sure." Dylan followed Taylor as she traversed the hallway maze and realized they weren't following the usual route. "Teaching me a secret way to the main lab?"

Taylor shook her head, "No, a different lab for different tests. Today we are using small-scale equipment that allow us to quickly evaluate a variety of test parameters.

"We set up the mini foam-filled tube in a small lab at the end of this hallway. We conduct tests on *the Scrubber*'s components in here."

Upon entering the lab, Dylan was immediately impressed with the complex array of analytical equipment and the high-energy activities of the lab staff. The tight timeline was taking its toll on the team. The staff knew that their funding grant was close to running out, and they needed a break-through very soon.

Hovering over the monitors and control panels, Professor Dave White nodded to Ethan. With a brief, wishful look upwards, as if completing a short prayer, Ethan began adjusting concentrations and flows rates.

Dylan again studied the team—each member had a lot riding on this test. Dave seemed edgy, unlike an objective scientist. When asked a question, his responses sounded tight and irritable. This test had to produce the data that proved the value of *the Scrubber*.

And then there was Ethan Rivers. What was with that kid? Maybe it was as simple as too much caffeine, but he was not unlike a small buck during mating season, jumpy and always looking over his shoulder.

The muted voices from Sully and Zhang huddled in the corner competed with the whirr of the monitors.

"Slowly increase the hydrogen and exhaust gas levels." Dave seemed somewhat pleased as he gazed at his tablet.

"Gas concentrations are increasing." Ethan's hands jumped across the dials. "Come on you little buggers. I know you can hear me! Eat that carbon dioxide." Ethan caressed the flow switches as if hoping to encourage the bacteria.

Taylor, excited, looked up from the readout as Dylan scooted next to her. Dylan inspected the monitors for the bacterial culture.

"Nice bacterial slurry, Taylor! Your parents must be proud," said Dylan.

Dave straightened up not as pleased as a few moments ago. "There's that damn variability again. It's as if the process is smooth and even, and then there is a big shift in results. I don't get it. Let's increase the carbon dioxide levels to protocol maximums."

Ethan's strong hands paused above the switches. "Shouldn't we take this in smaller steps?"

"Either the system is robust or it's not," barked Dave. "We need this performance data for the next phase of the experiment. Start ramping up the concentrations."

Ethan reluctantly began adjusting several knobs.

Dylan caught Taylor's attention. "Monitoring gas concentrations can be tricky if the gases aren't mixing evenly. Is there something else we can monitor to tell us how the bacteria are doing?"

"Well, yes, alcohol production," Taylor's attention cycled back and forth between Dylan and the dials. "The bacteria change carbon dioxide into alcohol. The more carbon dioxide consumed, the more alcohol produced. The alcohol won't kill the bacteria as long as the alcohol removal system is working."

Dylan nodded. Taylor's thinking made sense.

Ethan was uneasy. "Professor, I don't like stressing the system. Would you be OK with automating the concentration increases in smaller increments? Allowing a more gradual increase?"

"How much time are we talking here, Ethan?"

"Thirty minutes max."

"OK." Dave agreed through tight lips.

Ethan exhaled a big breath in relief. "Thanks, Dr. White. I'll very gradually increase the CO_2 levels over the next 30 minutes. It will take at least 20 minutes before we even get close to stressing the system."

Dave perched on a desk in the corner of the lab and thoughtfully bounced his tea bag in a forgotten mug of room temperature water.

Sully and Zhang, deciding this was an excellent time for a short break, headed for the lab door. Catching Taylor's eye, Dylan motioned towards the door. Taylor nodded but paused at the control panel resting her hand on Ethan's shoulder. As Ethan quickly acknowledged the encouragement, Taylor moved towards a gentlemanly Dylan holding the lab door open.

As Taylor approached the open door to the hallway, Dave cried out, "Combustible gas leak! Cut the hydrogen flow! We have a leak!"

Ethan's hands became a blur as they moved over the control panels. The hissing became louder and sounded more like a sustained shrieking.

"Everyone out!" yelled Dave. "This could blow!"

As Dylan turned towards Dave's voice, a force like invisible hands pushed him into the hall, throwing him to the floor and bouncing Taylor into the wall. A ball of flame rolled across the ceiling like a living thing trying to escape but was short-lived as the lab's sprinkler system engaged.

Dylan found his footing and reached into the lab to grab a crumpled, dripping Ethan. Stepping forward, Dylan grasped Dave's ash-covered lab coat pulling a ghost-like Dave out of the vapor into the hall.

The fading hissing sound was drowned out by the ARCS fire alarms. As Dylan helped Dave up from the floor, he quickly scanned the hallway. "Taylor, Check Ethan!"

Taylor winced as she moved her shoulder and gingerly crawled over to an unconscious Ethan. "Sully, get the first aid kit down the hall."

Taylor began checking Ethan's breathing and pulse, looking for any signs of trauma. As Sully skidded up next to Taylor with the first aid kit, the fire alarm stopped and Ethan appeared to regain consciousness.

Everyone breathed a sigh of relief. "He'll need to be checked, but I don't see anything serious," Taylor announced.

Dylan turned to Dave, "What just happened?"

"Damned hydrogen explosion! It must have been triggered by that gas leak. I heard some hissing as Ethan slowly increased the CO_2 flow rate." Dave wiped the water and ash from his face.

As his mind began considering the implications to the project, Dave looked sadly through the doorway into the lab wreckage. "Zhang? Are you OK?"

Only slightly disheveled, Zhang nodded affirmatively. Dave visually confirmed Zhang's response.

"Immediately start an investigation. I want to know what the hell happened."

"OK. I'll start now," said Zhang.

Taylor stood next to Dave, "Could someone have accidently left a control valve open? I don't see how that could have happened." *Would someone puncture the hydrogen line on purpose?* Taylor wondered.

27

GOVERNOR LEOLA BATES-HARDY sat at her comfortable desk and gazed at the box containing the model of *the Scrubber*. She pushed a button on her desk phone and asked her secretary, Shirley Ahn, to come in. Leola thought of her long-time secretary as somewhat of a partner since they'd been together for so many years.

Shirley walked in looking pretty as usual. Her mixed Swedish/Korean ancestry, slender figure and appropriate business attire made anyone who met her expect efficiency and intelligence. Her poise made her seem like she could handle any crazy situation that could come up in a governor's office, and she could.

"Shirley, in your mind, what's the mood in the campaign right now? I'm concerned about all the attack ads coming from the Super Pacs."

"Well Governor, you should be concerned. The latest polls show your lead narrowing. We have another poll coming in tomorrow, and everyone is really gloomy about what it will show."

Leola glanced over at the model. "I think I know how we can widen that lead."

"Well good. Yeung's campaign is saying all the things that the public wants to hear, and is painting you as a bleeding heart liberal who will spend too much on welfare, ban guns and ruin the economy."

"We need to do something about the environment, but to even mention that will scare away voters who think that to put responsible limits on the fossil fuels industry is to end life as we know it in Alaska." Leola stood and walked over the small refrigerator full of Dr. Pepper.

"Maybe so, but your campaign does not have the money to get the word out. You need to make sure your message is simple and consistent." Shirley looked disapprovingly as Leola opened a can of pop with a loud snapping sound.

"I want you to set up a meeting of the campaign staff and workers in the ballroom." Leola led Shirley out of the office and down a hall to the mansion's grand ballroom.

Leola's grandfather had scoured Europe buying up ceilings and walls to decorate his impressive home. The ballroom had the look of the Hall of Mirrors in the French palace of Versailles with ornate classical-themed murals, tons of gold leaf, candle chandeliers and stunningly large beveled mirrors to reflect the candlelight. Shirley's footfalls echoed in the huge room.

"I want my entire campaign staff here for the meeting along with all volunteers. Have the big screen monitor setup for a video call." Leola gestured to the end of the room where the musicians usually set up when there was a dance scheduled.

"Wait." Shirley looked startled. "You want everyone here? I suppose you want food, too?"

"Yes. Let's treat them well with an open bar, too. I need them fired up, and I think the phone call will do it."

"What's the phone call about?" Shirley wanted to know.

"It's about Alaska's resource-extraction economy, and what we're going to do about it." Leola took a drink from her Dr. Pepper.

28

PROFESSOR DAVE WHITE, his bald head covered in white ash, walked into the one place where he could separate himself from the world, his office. In this artificial cocoon he mentally focused on the critical issue of the moment by shutting out the routine day-to-day work problems and home life concerns.

An accidental glance at the family vacation picture on the wall momentarily reminded him of his pledge to his wife to use all of his accrued vacation time when *the Scrubber's* development phase was completed.

Sitting at his desk, he began to scan through the data from the lab. Luckily, the real time data had been routinely sent to the main lab computers throughout the experiment and data up to the exact moment of the tank malfunction, had been saved. Perhaps the information would at least provide some hint as to what the hell went wrong. That shouldn't have happened. This delay could ruin everything!

A knock on the door caused Dave to look up from his computer screen. Zhang opened the door and stood in doorway. "Professor, if not busy we show review and summary in conference room. In ten minutes. OK?"

"Thanks, I'll be there."

Zhang nodded and then abruptly departed.

Dave continued to scan *the Scrubber*'s operational data on several computer screens. The initial results had been encouraging. Carbon dioxide concentrations in the exit gas hovered around minimum detection levels at the exit sensors, significantly lower that the CO_2 levels at the input sensors.

These data suggested that the bacteria were efficiently transforming carbon dioxide and hydrogen gas into the building blocks needed to grow new bacteria and make alcohol. And just when Dave was thinking he might have cause for celebration, there was a significant jump in CO_2 levels. A moment later, the explosion interrupted the experiment and the data stream. *Shit!*

Frowning at the testing information, a shuffling sound caused Dave looked up to see Conrad standing in the doorway scratching his head with his prosthetic pincer. "We can go into the lab now. Let's assess the damage."

HOURS LATER a vigorous round table discussion started in the staff room. Ethan and Zhang had just completed their review of the explosion's cause, unexplained gas leak. The discussion moved on to test results.

Dave White scanned the team, assessing each member for signs of injury and stress. Ethan and Zhang were the most disheveled. He regretted that he'd put more pressure on his staff with yesterday's emotional outburst. Dave looked at Dylan, "So, Dylan, you've been quietly absorbing this discussion, what are your thoughts?"

"Well . . ." Dylan looked around at the eager faces of the development team. "From my vantage point, the science that we have so far makes sense. I only had a chance for a quick scan of the data, but I'm encouraged."

"OK, Dylan," Dave nodded "The good news in this tragedy is that we can easily change operations to the main lab, but our work is definitely slowed by this setback. We should be able to resume testing within six hours if we all work non-stop." Dave's expression was grim. He did not want to overwork his team, but grant deadlines forced his hand.

"The semi-bad news is that Governor Bates-Hardy wants a comprehensive report within 24 hours, and we have a scheduled teleconference in just a few minutes. I emailed her about this morning's problems." As Dave glanced around the room, he could see the reactions of his team varied from panic to resignation. The last response, belonging to Dylan, troubled Dave the most.

The "panic" responses made more sense. The visit of a funding source meant data must be organized into non-technical sound bites and taglines. The team did have a lot of work and extrapolation to do in a very short period of time. If the test had gone according to plan, they would have some very clear, but preliminary points to share with her. Since the test was interrupted, the data analysis was behind schedule.

"Dylan, you look uncomfortable," said Dave.

Dylan quickly examined each team member like he scanned a flock of birds trying to predict which one would leap to the sky first. "Frankly, I feel out of my element here. It's been a long time since I've done any work in a lab. And, I just haven't seen enough data to formulate any preliminary conclusions. As an outsider, I'm pretty much dependant on you guys for information. I'm sorry I'm not much help, Dave."

Dave stood up. "I wish I had a say in the timing, but unfortunately I don't. We need to prepare for the teleconference; Governor Bates-Hardy will be calling in about fifteen minutes. Let's take a break"

Dylan felt his heart rate jump. There was no way he could be expected to report on the status of *the Scrubber* project after only a 24-hour on-site visit. Not only would Hardy be hanging on every word, but also every nuance and innuendo would be immediately communicated to Anja. Given the difference between Hardy's and Anja's strategic objectives, Dylan's words could cause dramatically different reactions. He needed time to think.

"Dave, I'm not sure this is such a good idea." Dylan stood in the hallway outside the conference room. "I have only begun the process of coming up to speed on *the Scrubber*'s performance. I need more time."

"I'm afraid you are just going to call 'em as you see 'em." Dave needed a positive assessment from Dylan—there were no two ways about it. "How do you feel about the initial data?"

"That's like asking me, 'when you're in downtown Seward and take three steps toward Mt. Alice do you think it's going to be a rough climb?' The ascent toward Mt Alice begins on the other side of Resurrection Bay. There is no way I can feel any Mt Alice changes in elevation from downtown Steward."

"But you can see Mt. Alice in the distance."

"Sure, when it's not raining." This dialogue wasn't accomplishing anything. Dylan caved, "OK. I get it. I'll do the best I can."

The conference room display screen was glowing the "Welcome to the ARCS" text as Dylan and the team encircled the teleconference table. The screen indicated that the webcam perched on the monitor was active.

The video screen shifted to an empty governor's conference room. The timer in the bottom corner of the screen counted down from one minute. Dylan began oval breathing. Dylan realized how much he longed for the peacefulness of the High Rocks cliff near his cabin rather than the unfamiliar political precipice he now faced.

29

LIKE A TRUMPET slightly out of tune with the orchestra, the elegance of the Netsvetov Hotel sounded a discordant note against the small town feel of downtown Anchorage. Oblivious to the overstated elegance of the Netsvetov Hotel, Jane Koss bypassed the opulence of the hotel's reception area and ducked into the first elevator.

As the elevator moved her swiftly upwards, Jane studied her appearance in the mirrors and wished she didn't look so young. *Maybe if I got a boob-job, people would take me more seriously.*

Jane's thoughts drifted toward her office environment and Colonel William Bolton. He would stride through his office suite with an easy grace like he owned it. Located one floor below Bolton's living quarters, the office suite was home to the Super Pac.

Sometimes it seemed like Bolton wanted to move their brief relationship beyond its current *business-only* status; other times he made her feel like superfluous eye-candy. Jane shook off the thought.

The elevator opened onto a large, open foyer designed to impress any visitor. Jane turned toward the nearly hidden, modest corridor to her office.

If someone was beamed directly into Jane's office without any knowledge of the hotel or Bolton's adjacent offices, they would have seen nothing to distinguish her office from that of a thousand other senior executives.

Unlike the penthouse rooms upstairs, no evidence of gold or marble showed anywhere. The walls however, decorated with large, beautiful photographs of Alaskan landscapes and the furniture was handsome wood-and-chrome leaning more toward practicality than elegance.

Her cell phone chirped as she switched on her computer. Jane glanced at the text. *High marks for Bolton during the forum. I have more information for you.*

Looking at her phone, Jane recognized the unusual contact name, *Landslide*. This was the nickname for her secret contact in Governor Bates-Hardy's operation. Landslide had become an outstanding source of confidential information including Alaska Remote Climate Station cost overrun information and other wasteful spending by the Governor's team. Apparently some of the ARCS coal industry-funded experiments were yielding contradictory results.

Jane had vetted information supplied by Landslide in the past; it was always validated by other sources. It was as if someone in Hardy's operation believed in Bolton. Perhaps it was someone who was ready to switch to a new governor. Typically, when she passed Landslide's information through to Bolton, Jane quoted "inside sources". Bolton seemed curious but politely didn't press for specifics. Jane didn't want to speculate as to how Bolton thought Jane was compensating her source. Knowing Bolton's perspective, his conclusion would probably involve many horizontal nights.

Jane's attention returned to her cell phone as a second text appeared. An uneasy feeling crept over her as she read the text, *Anchorage governor's mansion, service entrance 7:30 p.m. Come alone.*

A chance to meet Landslide! The thought intrigued and concerned Jane. *Meet at the Service Entrance? What was this? Twenty-First Century cloak and dagger alive and well in Alaska?* There were alternatives, more understated venues for an off-the-grid meeting. Why the service entrance? Should she be concerned?

Jane wondered if she should mention this to Bolton. Although it's often good to keep one's boss in the know, Jane did not want to share any information about her secret source inside the Bates-Hardy campaign just yet. That was not something she wanted to do. *No.* Jane made the decision to protect Landslide's identity by attending the meeting without advising Bolton. Checking her watch, Jane noticed she had only twenty minutes.

30

DYLAN LOOKED LONGINGLY through the small conference room window to the edge of the snowy forest. The wind-tossed trees and blowing snow beckoned him to come outside and breathe the crisp forest air. *Not going to happen!*

Conference Room Two was empty except for Dylan. Both Professor White and the development team had expressed their concerns about being a distraction and had quietly exited.

Dylan felt a little like the last man standing. The conference room wall-mounted high-def screen gave Dylan the feeling that he was actually sitting in the governor's office, just the two of them, each representing their own agendas. Thinking back to his last meeting in the governor's office, Dylan could still feel his reactions, his double dose of surprise.

First, the resurrection of his college research project and secondly the admission by the Governor Bates-Hardy that she was becoming an environmentalist. And then, moments later, Dylan found himself streaking off to Alaska's Climate Research Station facility charged with representing both daughter Anja and re-election hopeful Bates-Hardy. *Those were quite the meetings!*

And now he was virtually sitting in front of Governor Leola Bates-Hardy as she teetered on the edge of her coming out–as-an-environmentalist press conference.

"Thanks for meeting with me, Dylan." She seemed comfortable and at ease as she cut to the chase, "What's your assessment of this project's progress so far?"

Dylan's controlled breathing helped him collect his thoughts and expel some of his tension. "Well, Governor, the preliminary results are encouraging. This research project is not unlike climbing Denali." Dylan didn't miss the governor's slight smile; she was obviously aware of Dylan's background.

"It's as if the ARCS team successfully left the base camp at 7,200 feet and was closing in on the Windy Corner when the lab explosion halted the experiments. Although the data was encouraging, Windy Corner is only at 13,500 feet. And a lot can happen during that last 7,000 feet to Denali's summit."

The governor's face was emotionless as she processed the analogy. "I understand that the team will be back on the trail, as you put it, within the six hours or sooner?" Dylan's nod encouraged the governor to continue.

"Here's my dilemma, Dylan. Greg Yeung is a strong contender for the governor's office. His campaign goes to great lengths to contrast his pro-oil stance with the *protect the environment/meet energy demands* tightrope that I walk.

"My platform announcements, due out tomorrow, will be timed and worded to support both Alaska's oil and coal energy resources without pissing off the environmentalists. The announcements will have a maximum impact if the timing is carefully controlled. Any stray or uncontrolled volleys from the fringe, and we're pretty much screwed."

Dylan knew exactly where Hardy was going. If the news of a successful *Scrubber* experiment were leaked to the press, the surprise and unifying elements of the plan would avalanche into oblivion. And, although the governor wasn't mentioning names, Dylan knew that Anja was potentially at the center of the avalanche.

"You know, Dylan, you are in a position to help me. You have, in our jargon, EC—Enviro-Credibility. I realize your research life ended a while ago, but we are not expecting a Master's Thesis.

Dylan could see where this was headed.

The governor continued, "And then, with your connections to Mountain Club resources . . ."

And there it was. A *personal* endorsement would go a long way to snagging the environmental voters—those undecided voters that Greg Yeung could never touch.

"Ma'am, I'm not sure I can help you out," Dylan said.

"Why do you say that? You are definitely an advocate for the environment."

"I am an advocate for the environment. I'm just not yet sure that *the Scrubber* is going to meet or exceed everyone's expectations. The data is too preliminary. It is too early for a public announcement."

"Who said anything about a public announcement?" Hardy seemed genuinely surprised.

"But I thought . . . "

"A public announcement is more the purview of Professor White. He heads up the research program. I was thinking something a little more modest and aligned with a technical staff position."

"You're offering me a technical position?"

"Actually, yes." Hardy paused thoughtfully. "Would you be interested in part-time position on my re-election team as a technical advisor?"

Dylan gaped, "Why me?"

"I'll be candid, Dylan. Four reasons. First, you are an expert on the initial science behind *the Scrubber*. Yes, Professor White and his team are moving the design toward practical reality. But you introduced us to the vision.

"Second, people can sense your independent spirit," the governor smiled. "Even if they haven't heard anything about you. We Alaskans highly value your attitude.

"And then there is your daughter. Anja is both a key player in the Mountain Club hierarchy and visible advocate of a better environment.

"Fourth, my campaign will benefit from a little reflected glow."

Dylan mulled over Bates-Hardy's words. He decided he did believe in doing whatever he could to help the environment. Even if all he did was to elevate environmental priorities in the public's mind—that in itself would be worth the effort. And what if these efforts helped him find a place in Anja's life?

"OK, Governor. I'll listen to the specifics concerning the path to your goals and then determine whether I can help."

"Thanks, Dylan." The governor appeared appreciative. "We'll get the employment paperwork started."

"So, in this capacity as technical advisor, how do you see me helping? Reviewing *the Scrubber*'s references in campaign materials? Simplifying *the Scrubber*'s science to the re-election team?"

"Exactly." The governor leaned into the camera giving the impression that she was about to share some confidential information. "Attacks from both the pro-oil faction and the environmental faction have been taking a big toll on our team's moral. The offensive maneuvering has been relentless.

"I need a way to give us a boost—some way to re-energize the team. And to heap more pressure on my team, a news conference is re-scheduled for tomorrow afternoon. I want a news media packet containing an overview of my re-election campaign ready for that news conference. I was hoping you might help us out."

Dylan began feeling a little uncomfortable. "I can't confirm a potential positive outcome for an incomplete project. And it sounds like you are asking me to do this.

Well Governor, I'd feel comfortable helping out under certain conditions. "

"Conditions?"

"Yes. This will need to be a private briefing, no reporters or public endorsement."

"Very well, I can comply. Your meeting is already scheduled for a very private location."

"Wait a second." Dylan assessed the implications. "How do you expect to get me to your team for a private meeting? Unless there is a helicopter waiting outside for me, there is no way I can get there in a timely manner. Secondly, how can you provide the media with the key parts of your re-election campaign when the lab won't be functional, let alone begin to collect data for hours?"

Governor Bates-Hardy shook her head. "Regarding the lab results, Professor White feels that there was enough data to draw preliminary conclusions. He and the team have been crunching the numbers all night. Professor Qi has suggested a retest of the bacteria slurry substituting a pure CO_2/hydrogen mixture for the exhaust gas from the combustion plant. He wants more control over the gas in those experiments where the data is not meeting expectations. Hardy held up her hand to delay Dylan's responses.

"Regarding the meeting, I'm sorry but I must not have been clear. Actually I would like you to meet with the re-election team via video conference."

"Oh. That might work. When were you thinking you'd schedule this meeting?"

"Well, how about right now. It turns out that the team is concluding another meeting next door and we're about ready to usher them in."

"What? I have no presentation prepared. In fact, I haven't even had time to mentally sort through my preliminary observations."

"Just provide the team with your initial thoughts."

"But . . ."

"Dylan, you are a common man who understands *the Scrubber*. Yes, I know that the prototype has evolved significantly from your original work. But the biochemistry is the same. It's like a different mountain peak but you still use the same basic climbing skills." Dylan wasn't able to stop his grinning response to her surprise metaphor.

Governor Bates-Hardy acknowledged her sincere appreciation as the team filed in. Dylan took a sip of water from the cup on the conference table and tried to plan his opening remarks.

"Mr. Baker?" Dylan looked up the see the re-election team standing in the ornate ballroom of the Anchorage governor's mansion.

"Thank you very much for joining us, Mr. Baker." Harold Cheek, the governor's senior political advisor, seemed a little full of himself. "Governor Bates-Hardy mentioned that you had some information for us."

31

ZHANG QI HAD GROWN up in Taiwan, the son of a successful business family. His parents had started manufacturing welding gloves for the US market and the money had started pouring in. When Zhang was born, his parents expected him to grow up and take over their thriving business, but Zhang wanted something different.

A goal of attaining an engineering degree from an Ivy League had him applying at all eight of the prestigious colleges, but was only accepted at his back-up school, the University of Washington. It turned out to be a wonderful fit for Zhang. His professors saw a driven student who had a knack for finding unique solutions to challenging engineering problems.

Partway through his engineering degree, Zhang discovered a strong passion for organic chemistry. His enthusiasm and curiosity for applying engineering to chemical problems allowed him to be part of the teams that discovered breakthroughs on carbon dioxide metabolism studies at the University of Oregon.

Despite his academic successes, his parents wanted him to return to Taiwan and become a businessman. He wondered what he would have to do to prove to his family that he had made the right decisions with his life.

ZHANG'S CURIOSITY OFTEN TOOK him places other scientists never thought to go. Right now he was pulling one of the Layer Cake Membrane filters in Main Lab 4. Dave White usually had techs do this work, but Zhang wanted to see these miracle worker germs for himself.

The first thing that he noticed was a strong alcohol odor. *That's not right, the alcohol levels in the filters should be extremely low because we drain it away.* He opened one of the filters to see a black oozing goo instead of the healthy, white smear of a viable bacterial culture.

Something killed our bugs. He retrieved a specimen container and scraped some of the goo into it. All the time he wondered what could have caused the culture to die. Of course his first thought was that the alcohol produced by the culture had not been moved through the tubes and down to the collection basin. He'd need to check the basin near the filters to see how much alcohol they produced in the last test run.

Zhang also thought about the coal being burned off-site. Impure coal containing high levels of certain heavy metals like arsenic and chromium could kill the culture.

Dave had assured him that there would be no contaminated coal introduced at the combustion station. He knew they had sealed the burn chamber according to proper protocol, but sometimes things go wrong. He'd seen it before. At a test site in Berkeley, someone had accidently dropped a contact lens in the combustion chamber causing an amazing amount of bad data.

He felt excitement at the thought of discovering and solving this problem for the team. Sometimes he felt isolated from the others due to his English skills. He knew there were sounds in the English language that tripped up the Chinese brain. And the English language itself made no sense. Who makes up a language with 13 ways to spell the /sh/ sound?

Since the weather had calmed down, he could dash out to the combustion station, check out the situation and be back in an hour. If he were caught breaking the safety protocols about leaving ARCS without a partner, he could claim he didn't understand the rules.

Zhang loved technical mysteries. This mystery had two red flags, alcohol and black goo—neither should be present. He hurried to the lab to borrow a nuclear magnetic resonance spectrometer. He couldn't find one anywhere, so he pried open a drawer with a sign that said *Sully's Stuff. Do not borrow!* and grabbed Sully's Hytech Ms.2921a. Then he rushed to the bunkroom to pick up his bright orange outdoor gear. Once suited up, he let himself out when he thought no one was looking.

A SLENDER BEARDED MAN wearing a kitchen white uniform observed Zhang's departure and quickly moved to report Zhang's unexpected actions.

32

DYLAN BAKER STARED at the image on the ARCS Conference Room high definition video screen and finished another round of controlled breathing. He could see a picture of what was happening at the governor's end of the call, also in the lower corner of the screen, a very small image of an uncomfortable man with a bad hair day: Dylan Baker.

The teleconference terminal at Governor Bates-Hardy's office appeared to be in a ballroom where the re-election team had gathered and were waiting patiently for Dylan to begin. *How did I get myself into this situation?* Dylan wondered. Governor Bates-Hardy had been extremely persuasive. *OK, just tell them what I know to be true.*

"As you may know, I'm an outdoor guide and handy man," Dylan paused to survey the team standing in the huge ballroom in Anchorage. The governor and Senior Political Advisor Harold Cheek took up positions on the periphery of the room holding drinks and a paper plates piled with appetizers. Dylan initiated a slow inhale, pause and slow exhale. He was sure they heard him take that breath. The teleconference connection was excellent.

"Knowing very well my research history, a persuasive Governor Bates-Hardy offered me an all-expense paid trip to the Alaska Remote Climate Station if I would report back on the status of *the Scrubber* project." Several people around the conference table smiled. Others nodded acknowledgement of the governor's negotiation skills. Dylan cycled his breathing again.

"It's a high energy day here at ARCS. Outside, Mother Nature is swirling the snow, while inside, the development team pours their hearts and souls into this research project."

Dylan noted the verve around the re-election team's conference table seemed to be ebbing away. People were starting to glance at cell phones and randomly look around the conference room.

Dylan glanced quickly to the window framing the forest surrounding the ARCS. *The hell with it.* As his best friend Shorty used to say, "Always fight back if attacked by a black bear." Dylan squared his shoulders and looked directly into the conference room camera.

"OK, here's the update. Bear with me. The results are still a little preliminary. While you've been preparing for the campaign, the governor has been supporting a project to resurrect and update some previous research data.

"My college research demonstrated a method for capturing and scrubbing the carbon dioxide out of combustion exhaust gases. Other research during the past decade demonstrated the possibility of small-scale conversion of exhaust CO_2 to a gasoline additive. Although too soon for a valid conclusion, I find the preliminary results for scrubbing CO_2 from coal plant emissions and converting it to alcohol very encouraging."

Dylan could see the team's reaction. The powerful ramifications of a new energy source with a minimal eco-footprint and access to previously unavailable, eco-friendly voters jumped from team member to team member. The team exchanged looks ranging from cautious optimism to giddy enthusiasm.

"The implications for this research go way beyond the environmental impact of coal-based energy plants to everyday life. Scrubber technology is Dr. White's program that supports the growing bacteria that pull carbon dioxide from exhaust gases. The Scrubber technology could be applicable to any mechanical system that generates CO_2, whether electrical, thermal, transportation or industrial source."

Dylan could sense the increasing pulse of the team. "As you know, Governor Bates-Hardy is planning a major announcement next week. Within the context of a State of the State address, she will issue a brief synopsis of the latest ARCS research results." Dylan turned his attention to the governor. "Is that a fair summary, Governor?"

Governor Bates-Hardy, pleased with the team's reaction to Dylan's brief summary, nodded affirmatively.

"Excellent." Dylan felt some of the presentation pressure easing. "Well then, first I'll take a few minutes to summarize my college work. After that, I'll provide a very high-level look at the objectives for the ARCS project. And then, I'll be happy to try to answer any of the many questions about my first impression of *the Scrubber*."

33

SHARP GUSTS OF WIND blew snow and ice particles around two tall high-tech towers. A figure wearing an orange parka and carrying a compact testing device, stopped at the first tower, retrieved something from a toolbox and, holding a flashlight in his mouth, started opening a shiny black panel on the side of the tower. Swirling snow made it hard to tell what the figure was doing.

A half-mile away in the warm Zetros communication room, Tech Sergeant Denver took off his headphones and approached Tango's private room. He tapped lightly on Tango's door. "Ma'am, I think you should see this. There's activity around the combustion station. You said to call you if I saw something out of the ordinary."

Tango opened her door wearing black panties and a sports bra. Completely unembarrassed, she led Denver to the communications room and peered at the monitor screens. Denver stood behind her admiring her athletic figure.

"Sergeant, stand down. I can see you ogling my ass in the monitor's reflection. I need you to focus on your job. Now what are we looking at?" Tango glanced down at the complex array of switches linked to the sets of cameras and mics in and out of the ARCS.

"This is a view of the combustion station. It's where they test and burn combustible materials, typically different types of coal. They then bottle the exhaust gases for use in their CO_2 scrubbing experiments."

Sergeant Denver watched her click through all the cameras to see if there was anything else amiss. "What have you heard from the interior mics?"

"Nothing about this, Ma'am. Only expected chatter has been picked up. I'm guessing this person is conducting an independent investigation."

"If he finds something and reports it, it will be too late to stop the leak."

The sergeant looked nervous and could only choke out, "Yes, Ma'am. What are your orders?"

"Contact Concord. Tell him to find out who that is and to place a priority one surveillance protocol on him." She looked at Denver. "This may be our first kill here at the ARCS and it needs to look like an accident.

34

SUNSET WAS STILL hours away as Jane Koss drove her Suzuki X90 along the gravel road to the back of the Anchorage Governor's Mansion. When she moved to Alaska, her father had told her to buy a 4-wheel drive car. When he found out she'd purchased a 1998 Suzuki two-seater, he'd told her she was a dumb ass.

The car had cost her much more than she expected. It was a former Red Bull advertisement, so she had to have a giant can of Red Bull taken off the car and have the car repainted. Then the fuel pump went out, twice.

Still, she loved the 4-speed manual transmission and the amazing six-CD stereo. In the soft spring evening, she had taken off the T-top panels and the car was as close to a convertible as she ever wanted to get.

Jane saw a state police checkpoint as she approached the service entrance for the mansion. Nervous at meeting her source, Jane took a deep breath and slowly exhaled. Why was she so uneasy about driving up to the gate and talking to a trooper? Was something telling her this meeting was a trap? Would she be arrested?

The bored female state trooper seemed glad to see someone come down the service road behind the mansion. At this time of the evening, things were pretty slow. When the trooper checked Jane's ID against the list of approved visitors, Jane was promptly waved through as if she were a VIP.

It was only a meeting with someone connected to the governor's reelection staff, but it had to be someone with enough clout to get her through the first level of security so quickly.

Jane felt her anxiety grow. She tried to tell herself not to worry; it was not like anything bad was going to happen.

The back of the mansion had none of the luxurious green lawns and blooming formal gardens of the front. It was mostly parking lot, gravel and bark chips, but the sparkling windows of the famous Victorian mansion seemed to reflect away some of the stress that emanated from Jane.

She approached a loading dock labeled with a *Service Entrance* sign. To the side was a sturdy metal door with a call button and security camera. The metal door looked a little intimidating, more like a barrier than a welcoming portal. Jane glanced at her watch. She was ten minutes late. *Well, I hope a few minutes won't hurt*, she thought. Before she could push the call button, a voice came through the speaker.

"Jane Koss?"

"Yes, it's me." Jane found her voice had gone up an octave. The door buzzed then opened. A small trim, well-groomed man regarded her condescendingly. He looked like he used some kind of greasy hair product from the 70s to hold a razor-sharp part in his dark hair.

"Jane Koss? We were expecting . . ."

"An adult? I'm 30 years old," Jane allowed irritation to creep into her voice.

"Of course. Sorry. I'm Phil Small." He mechanically extended his hand to her.

She shook it tentatively. "Nice to meet you."

"May I see your ID please?"

Jane showed Phil Small her Texas driver's license, which he studied under a hand-held UV light.

He stepped back into the dim light of the entryway, and Jane took the cue to follow. "Welcome to the governor's mansion. I was asked to escort you to conference room D. It should only be a couple of minutes." Phil Small held himself erectly and took quick, confidant steps as he led Jane through a maze of plain, dark hallways that could only be described as servant's quarters. "Would you like water or coffee? All we have is instant coffee at this time of the evening."

"No thanks." Jane followed him to a plain metal door. He put his hand against a sensor and the door slid open revealing a lavishly decorated hallway. Jane felt awed by the classic Victorian decorations, which seemed tasteful and genuine to her whereas Bolton's penthouse seemed merely gaudy and trite.

Mr. Small stopped by an elegant mahogany door subtly labeled, Conference Room D. He swiped his ID card in a reader concealed near the polished brass door knob, and opened the door to reveal a modern conference room that looked like it was imported from a thriving New York law firm, lots of oak, leather and glass. "Please wait here."

He gestured to easy chair in the corner. Mr. Small backed out of the room, closing the door and leaving Jane to review chic frameless photos printed on metal sheets.

The photo collection's themes were obvious: wild Alaskan flora and fauna, and landscapes of Alaska at dawn and dusk. Jane admired one that showed a climber roped up on Denali during a fierce storm. Jane felt as if she was in a similar storm, but she had no rope.

35

THE FIRST THING that surprised Zhang when he arrived at the combustion station was that the access panels were not locked. Anyone with a screwdriver could open them. For the first time, Zhang began to wonder if someone was sabotaging the experiment. The idea seemed preposterous. Who could get to this remote area without that someone noticing or leaving a trace?

Once he got the panels open, he noticed warm air flowing up from the base of the combustion station. That was strange. There should be no heating equipment out here, and the very last burn was several days earlier. Then Zhang plugged the testing meter into the leads so he could download the raw data and send it to the ARCS server. While the information was downloading, he checked the coal storage bin.

Zhang thought the coal looked like the nasty stuff his family burned when he was a kid. Dave had told the team they were testing some ultra pure coal, but this batch looked like the dirty, brown coal characteristic of this part of Alaska.

The testing meter beeped as it started the download. At ARCS, the data from the experiment had already been processed and loaded onto spreadsheets and graphs for Zhang and the rest of the team to analyze. By getting the raw data from the combustion station, he could see the unprocessed information and compare it to what showed up on the ARCS computers.

Zhang could look for patterns and anomalies to help explain the presence of alcohol in the filters or the bacterial die-off. Since the combustion station was a port where all the unprocessed, raw data from *the*

Scrubber's tests were stored, all the information being loaded onto Zhang's testing meter would be unaltered. Zhang expected to find transmission errors that would explain the variance in the ARCS data.

While the download was proceeding, Zhang looked around. The sky threatened more snow, but the winds were calm. Once the snow stopped blowing, this part of Alaska was really a beautiful place. He could make out a forest to the west, but up on the ridge where the combustion station was located, there wasn't much cover. That's probably why the combustion station was built in this spot.

Zhang was afraid of bears. Cognitively, he knew the danger was quite low. Bear attacks were very rare and even less common this time of the year. Emotionally, he worried that a huge grizzly would burst out of the woods and attack him. It would be a horrible death, watching his flesh go down the throat of a monstrous beast. He looked around nervously.

When his data reader beeped, indicating the download was complete, the loud noise made Zhang's heart skip. Zhang removed the wire leads of the testing meter. He wanted to check his testing device before he replaced the screws.

A gust of wind came up and reminded Zhang that he should go back to ARCS now, but instead, he drew his hood up and started comparing the two data sets right on the testing meter. Sully had bought the good meter that would allow a user to compare two data sets in various ways.

What Zhang found almost made him drop the meter. Something was seriously wrong. What popped out of the raw data was a rapid change in efficiency as the coal burn continued. The CO_2 scrubbing process, where bacteria pulled CO_2 out of the air and funneled it into their cells for growth, started out with amazingly efficient results—far better that the ARCS data revealed, but rapidly declined.

The biocatalytic converter's performance data must have been averaged before it got to their computers at the ARCS station. This would mean that the drop off wasn't noticed. *Why would the data be averaged? Don't we need to know the converter's efficiency at each moment throughout the run?*

A loud voice made Zhang jump. "Hey! Who you?" Zhang turned around to see two polar bears rushing at him. He screamed.

Then he realized the bears were, in fact, two men in white snowsuits rushing up to him. One of the men looked familiar to Zhang. The other seemed to hide his face inside his hood, and he carried a heavy two-gallon vacuum bottle from the lab.

"You scare me." Zhang said. "I thought you bear, but you cook. Good cook. You are Josh."

"You need to come with us. This area is off limits to ARCS staff." Josh said taking Zhang's arm and pushing him back. Zhang tripped and fell onto his back.

"Hey. I on ARCS science staff. I help build this." Zhang was getting mad now. The second man had set down the large bottle and started to unscrew the stopper.

"What you doing that? That in we carry liquid nitrogen. That out here not belong. Only for lab use."

Now fear was bubbling up in Zhang's mind. It looked like Josh and his buddy were going to pour the super-cold liquid onto Zhang. The first thing that popped into Zhang's mind was a demonstration of the liquid when his undergraduate professor had dipped a rose into a beaker of the misting liquid nitrogen and then dropped the rose. It shattered into tiny pieces with a sound like breaking glass.

Suddenly Zhang's face was on fire as the misting liquid hit his skin and immediately boiled. Josh poured it in a strong stream over Zhang's head and filled his hood with an evil sizzling sound as the liquid nitro rapidly changed into a gas. Zhang lost consciousness nearly instantly as his head transitioned to -341 degrees Fahrenheit.

"OK, let's bury him in snow so it looks like he died of exposure. Don't touch his head, it will break." The two men carefully dragged the misting corpse near a snow bank and covered him.

A sharp wind blew more snow over the icy grave. Shivering, the men started back to ARCS carrying the empty vacuum bottle. Somewhere in Zhang's icy grave, the testing meter beeped: *data upload complete.*

36

WHEN TAYLOR FIRST ENROLLED at University of Wisconsin, her goal was to be a firefighter. Rebelling against her conservative Midwest upbringing, she wanted to attend a party school and afterwards get a job where she could help people and work *48-on/24-off* with a group of fit, lusty men.

She enrolled in the paramedic program, became politically active and discovered ultimate frisbee. Her team, Bella Donna, went to nationals the first year she played. At about that time, she fell in love with the men's ultimate team. Being pretty, smart and athletic, she got to take her pick of the amazing male athletes until she met Dan.

Besides sports and girls, Dan had a love for microbiology. He drew Taylor into biology with all its complex rules and puzzles. She quickly lost interest in Dan and focused on her studies.

For some reason, she developed an obsession with the anabolic flexibility of different bacteria to utilize carbon as a source of energy. Before she knew it, she had her doctorate in microbiology and two national ultimate frisbee titles.

She was casting about trying to decide whether she should go to work for some place like 3M or develop an academic career, when Dave White read some of her papers and asked her to join the ARCS team. The idea of being stationed on a remote base with lots of rugged men, plus having a chance to save the world appealed to her.

Then she met the men at ARCS. The smart, athletic ones were few. Ethan looked too pretty to be straight. Dylan appeared to be the best of the lot, but he was married and seemed loyal to his wife. Still, she could wait to see what developed. Sometimes people got bored and horny. Something fun might happen.

SHE ENTERED THE BREAK ROOM to see Conrad, Sully and Dylan hunched over a tablet. Sully looked at Taylor, "Hey Taylor, look what I found."

Pages of data covered the desktop and all horizontal surfaces. "What am I looking at?" she asked.

"Sully pointed to a data line with a corndog, "Why did the carbon dioxide take a jump here?"

Taylor looked at the graph. "A CO_2 spike appeared just before the exhaust gases stopped flowing into the Scrubber. The spike looked too rapid to make sense."

"It's abrupt, like the bacterial mat was somehow turned off or removed," said Conrad. "Can it be a surge in CO_2 production in the coal ignition?"

Taylor's smile vanished, "What the hell? Where did you find this? It wasn't on my data set."

"I think my tablet captured a raw data stream from the combustion station. I was just fooling around with my tablet as the data started coming in, and this popped up," Sully beamed.

"Actually, that's my tablet." Conrad scratched his head with his hook, "This seems to be an anomaly."

"Anomaly, hell. It's impossible," said Taylor.

37

ETHAN CARRIED A LAPTOP across the hall into a lab and over to the bench where Conrad was working on a software project. "Ah, Conrad? Don't stand up, I'm just here to ask a favor."

"I am standing up. I'm just real short," Conrad joked from his perch on a gray metal stool. Ethan found himself somewhat uncomfortable with Conrad's self-deprecating humor, but decided to ignore it.

"This is Sully's laptop. He spilled beer on mine, so I'm using his. I'm a chemist. When my laptop freezes, I try the normal fixes like restarting it, checking for software updates and so forth. Then I take it in to get fixed. Obviously I can't take this in to get fixed while out here in the field." He held up a sleek aluminum laptop.

"So I take it his computer is freezing, like everyone's ass out here near a mountain in Alaska?" Conrad took the machine and put it on his desk. "I'll take a look at it later."

"I need it now." Ethan looked sheepish. "Look, I can give you this bag of chocolate-covered mints from Sully's own personal stash." Ethan handed Conrad a bag decorated with scarecrows and zombies.

"Holy Cow, Sully eats these? That's leftover from Halloween! Don't eat it, put it in a museum. I thought diabetics were not supposed to eat a bunch of candy?"

"Sully has a candy problem," explained Ethan. "It's what he does best."

"OK. Let me see that computer. Actually, everyone's computer has been crashing, so I should figure this out now. Come back in an hour." Conrad shoved aside the project on which he was working, took out a portable hard drive and plugged it into Sully's computer.

"If it's ok with you, I'll just watch. I want to see how this is done." Ethan popped a mint into his mouth and took a position behind Conrad.

"You can watch, but stop eating those mints around me. First thing I'm going to do is to delete the porn and the bootleg games Sully put on this laptop. Lots of bugs come in when people put sketchy things on their computers." Conrad's one good hand moved with amazing speed over the keyboard.

"Next, I'm going to run this diagnostic program that I wrote." Conrad started up a program titled *Left Hand of God*. The men watched a progress bar slowly move across the screen. His prosthetic hooked pincers twitched.

"Hey Conrad, what ever happened to your hand? I hope you don't mind if I ask." Ethan spit out the mint and made a face.

"Nah. I don't mind. I was born this way. My left hand is perfectly normal."

"No, I mean what happened to your right hand?"

"Oh that? Why didn't you say you wanted to know about that hand? Same thing. Born that way, no right hand. Also missing some leg and spinal bones, too. I figure that somewhere someone has an extra hand and some of my bones."

The computer made a loud fart sound and a report started scrolling down the screen. Conrad examined the report with an intensity that he often hid with his humor.

"Whoa. There's some major code snippets in your analytic software that don't appear to belong." Conrad pointed to the screen. "These snippets look like filters designed to hide or discard certain data."

"I swear. I didn't get pirated software. This software came from Professor White's office. All of us have the same program. We use it to collect reports from the lab and turn the data into spreadsheets and graphs." Ethan spoke between absentminded bites of another mint.

Conrad inserted another hard drive into the laptop and scrolled through the contents.

"He gave that software to me, too. It came with a package of stuff when I was hired. Let's have a look at the backup I made from the original." Conrad opened his own laptop and checked the software on its hard drive. "Here it is. Same software as you have, but my program is much smaller. I bet yours became corrupted somehow."

Conrad looked surprised, and took a mint from Sully's stash. "Look. My copy of Dave's analytic software has been modified. It must have happened here at ARCS over our little network."

"Then why isn't your computer freezing?" Sully asked.

"I don't use that software. I haven't even opened it since we got here."

"Then is someone trying to alter test results?"

"No way. I wonder if it's happening by accident, and why am I eating this crappy mint?" Conrad spit it into a trashcan. "I don't know, but I'll write a little routine that will remove these nasty snippets and keep our software more secure. Tell the team to check their emails for the patch."

SERGEANT DENVER ROLLED his chair away from the COM center in the Zetros, "Ma'am. We have a problem. They discovered the software's hidden actions somehow and they're going to disable it. I think they suspect something funny."

"Shit! Arrange a call to The Boss. This wasn't supposed to happen." Tango paced in the small office.

JANE PACED IN THE IMMACULATE conference room. She hated this waiting. Waiting to meet her source. Who would it be? Would it be someone she had heard of? It seemed like the person was not a low level staffer since other staff members seemed to be at her source's beck and call. Up until now she thought it might a campaign volunteer.

Jane stopped in front of an amazing photograph printed on metal. It glowed as if lit from behind and the details were so sharp, that it compelled her to stand close to the image and explore even the corners. It was a picture of Governor Bates-Hardy hang gliding at sunset near the Alyeska resort.

A loud click from the door startled her. Her source?

Phil Small, looking like an overdressed secret service agent, stepped in. "Since you came through the service entrance, you bypassed security. You need to go through the scanners."

"The scanners?" Jane said weakly.

"We need to make sure you aren't carrying a firearm, explosives or recording equipment. We have a political campaign going on and need to take precautions." Phil Small spoke as if it were obvious that someone so tiny and slender could be hiding a weapon.

Jane was instantly glad she hadn't brought her digital voice recorder. If she wanted to record something, she could still use her phone, unless they took it away.

She followed the man's shiny shoes through endless elegant rooms and hallways rich with blond oak wainscoting and Victorian-style lighting and art. "We're going the long way. Since you are part of Greg Yeung's Super Pac, we're going to skip going through Governor Bates-Hardy's campaign offices. Too many uncovered white-boards with strategic planning notes."

Soon they came to a set of impressive doors leading to the front entrance of the mansion. Jane hardly had time to admire the beveled glass ornamentation on the doors, when they stepped outside into the fragrant, cool evening air. They crunched over a pink gravel walkway to the side of the building where the security screenings took place.

Jane felt as if she were in an empty airport screening area. She took a plastic tub and put in her shoes, purse, keys and phone, then stood in a booth with her hands over her head while state troopers gazed at a monitor.

Once out of the booth, a dimpled, fresh-faced female trooper asked her to stand still while she passed a wand over Jane's back. "We picked up some metal back here." the trooper apologized.

"It's probably the clasp on the back of my bra," said Jane. To her own ears, she sounded guilty of something.

Instead of picking up her purse and other belongings, another trooper handed her a receipt for the items and a set of slippers. Phil Small pointed to the receipt, "You'll get your things after the meeting. The person you are meeting with does not want you to have any recording devices with you."

After another walk among the mansion's oriental carpets, stained-glass lampshades and elaborately decorated ceilings, Jane and Phil stepped through yet another set of impressive doors to the mansion's visitor's office. Like most people, Jane was dazzled by the grandness of the huge ornately carved wooden desk, the leather smells and the dramatic lighting, some of which was coming from sunset-slanted rays shining under the heavy window drapes.

A forgotten can of Dr. Pepper sat on a side table near the window. "Please sit here. Your meeting will start in a few moments," Phil said gesturing to an upholstered upright chair near the massive desk.

As she sat in the uncomfortable chair, Jane could feel her heart beat faster. Her source was not just some staffer in a cramped office. Her source had to have some real power, maybe even the most power. She didn't dare to think who had that kind of clout in the governor's reelection campaign.

A door opened near a painting of a naked Rubenesque woman and out stepped Harold Cheek, grinning coldly.

"Good evening Ms. Koss. So nice of you to come and talk with me."

"But you're Harold Cheek." Jane blurted out, then regretted saying it. This had to be some kind of trap.

"Yes, and I am Landslide."

39

DYLAN ARRIVED in the lab and peeked over Sully's shoulder. Taylor picked up the tablet taking it away from the views of Conrad, Sully and Dylan. She glared at the screen, and then started poking at it.

"Hey, that's my tablet, " whined Sully.

"Actually it's mine," said Conrad looking up from his computer display. "You borrowed it because you spilled soda on yours."

"Where did you get these data?" said Taylor. "This is not what my bacteria are supposed to do. I expected to see much higher cell densities in the spent media."

Dylan looked blankly at Taylor.

Sully went on, "OK, when the data was coming in, I was using my tablet to try and jump to the front of the line. I wanted to see it first and not wait for it to come in on my laptop."

"So you grabbed incoming data with your tablet?" asked Dylan.

"Yep, and this baby has some kickass software on it." Sully patted the tablet.

"That's my software," said Conrad.

"Why are we squinting at a tablet? Let's put this up on a big screen and figure it out," suggested Sully. "We can use Conrad's big monitor."

"Please, no drinks by my computer," Conrad pointed to the can of soda in Sully's hand.

The three scientists and Dylan left the cafeteria and headed for Conrad's workstation. Dylan followed behind.

As they walked down the hall, Dylan wondered what he was doing with all these brilliant scientists. He questioned his ability to contribute to their discussion of carbon metabolization. His primitive method used a vented lab test tube, some soil bacteria and bottled CO_2. What did he know about alcohol synthesis beyond the Wikipedia summary?

Dylan looked at Taylor's feminine athletic form ahead of him. It made him miss Suzie. He wanted to ravish his wife and hold her as she slept. He shook his head. *This is not what I need to be thinking about.*

Once in the lab, Sully rushed to Conrad's chair and turned to him, "Conrad, what's your password?"

"Why should I tell you my password? You'll borrow my computer?" Conrad said.

"Come on, you can always change it when we are done here." Sully begged.

"OK. It's *pi* out to twelve digits after removing the '3.14'," confessed Conrad.

"Oh, I should have guessed that, " said Sully. "So it's 159265359?"

Conrad looked glum, "Yes."

"That's a good one. I was going to guess it was the half-life of carbon 14 times the square root of negative one."

"There's no such number," said Conrad.

"If you had better software on your tablet, you could figure it out," said Sully.

"Sully," said Taylor. "Just put those data up on the screen so we can figure out what's going on."

Sully approached the keyboard. An array of data sets, arranged in folders, came up on the large monitor. "Hey, none of these folders have the code for the set on my tablet," his gesturing hands knocked Conrad's mouse off the table. It hit the floor and burst open sending batteries rolling under a shelf.

"That's my tablet," said Conrad looking concerned. "Now you broke my mouse."

"Give it here. I want to see the address of that chart we are looking at." Sully pulled the tablet out of Taylor's hand. He glanced down at the mouse, and looked up at Conrad. "Sorry. I'd give you my mouse, but its batteries are dead."

A big voice from the door boomed into the computer lab, "What's up? Did Sully wreck the damn microwave again?" Dave White walked in.

"He broke my mouse," said Conrad, his whiny voice sounding like he was tattling.

"That wasn't my fault. How was I supposed to know there was metal in my lunch when I put it in the microwave?" Sully said.

"Well, it was your metal spoon in the bag," pointed out Dave White. "Really, what are you guys working on?"

"Sully's tablet is displaying different data than what came through on our computers," said Taylor. "It looks like a major drop in the spent media cell count." She pulled the tablet from Sully's hands and gave it to Dave.

"That's my tablet," said Conrad his voice rising.

Dave scowled at the screen and looked concerned. "Whoa. We need to get to the bottom of this. The bacteria in *the Scrubber* are in exponential growth. The spent media should have contained very high levels of bacteria. What do you think it is?"

"I can't find the data that's on the tablet anywhere in these folders," said Sully as he clicked through folder after folder. "Let me see my tablet again. I'll do a search for the exact name of the files."

"That's my tablet," persisted Conrad. His mind went back to middle school when some big boys grabbed his book bag and played Keep Away by throwing it back an forth just out of Conrad's reach.

"Jeez, Conrad. You're so possessive. Lighten up," said Sully as he copied the file name character by character into a search field.

"Lighten up?" asked Conrad. "What do you mean, *lighten up*?"

"It means Sully thinks it's his tablet now," said Dylan.

"Crap! That file is nowhere in any of the data sets. It only exists on my tablet," exclaimed Sully.

"Sully, let me have your tablet," said Dave. "We need to get to the bottom of this. Let's hope some data just got scrambled. There should be no filtering of results to my scientists."

"My tablet," moaned Conrad.

Dylan seized the tablet out of Dave's hands and handed it to Conrad. He turned to Dave, "Ask Conrad if you can borrow his tablet."

Surprised, Dave cast a sheepish *I'm sorry* look to Conrad.

40

JANE HAD NEVER SEEN Cheek in person. It was clear that photos and TV flattered him. They did not reveal acne scars poorly hidden by pancake makeup, yellowed teeth, greasy hair or the nervous tic Cheek seemed unable to control except on camera. He would repeatedly turn his head sharply to the side as if to crack his neck.

"So, welcome to the governor's office," Cheek bent forward and offered his hand. She could smell a sour body odor coming from his direction. Inwardly she found him creepy and repulsive, but shook his limp, moist hand.

"You're my source? You're my spy in the governor's office?" Jane asked.

"Why yes, I'm Landslide." Cheek ticked his head and sat back in the big chair behind the desk. Jane realized the meeting place had been staged to throw her off balance. Her chair was short and uncomfortable, his was grand and allowed him to look down at her. She noticed he had long white hairs protruding from his nostrils and mixing with his mustache hairs.

"Why have you been helping our 501? Everything you have sent me has helped Greg Yeung and damaged the governor." Jane tried to take the lead in the meeting.

"If you say so. I'll admit that some of what I sent you has caused some temporary setbacks in our campaign. However, our goal has been to force Yeung into making his energy position something from which he can not change." As he talked, Cheek absentmindedly cracked his knuckles joint by joint. The noise was startlingly loud.

Jane became even more on her guard, "Why would you want to do that? It has only made the governor look more like a bleeding heart environmentalist?"

"Exactly so, Ms. Koss. Do you realize that 97 percent of climate scientists have agreed that the activities of humans have caused global warming trends? Did you know that it's a proven fact that our increased carbon dioxide levels are warming the planet?"

Jane made no response to Cheek and this seemed to add vigor to his speech.

"Global sea changes include a rise of 6.7 inches over the last 100 years. The sea is now 30% more acidic than before, and Alaska's snow cover is shrinking faster than our glaciers are retreating." Cheek head twitched to the side with greater frequency as his voice rose.

Jane continued to gaze emotionless at Cheek. She was concerned about these things, but at this moment in time, she had to look out for her boss's interests.

"And that asshole Yeung wants to block alternative energy subsidies and develop more carbon-based energy sources. You're damn right the governor comes off as an environmentalist. The voters' choices will be black and white once you and I are done working together."

Little bits of spit and big puffs of whisky/coffee bad breath were being cast in Jane's direction. She tried sitting back, but the knobby brocade of the uncomfortable chair sent little pricks of pain into her back as she leaned away from Cheek.

"Mr. Cheek, I'm not going to sit here and be yelled at. Why did you call this meeting?" Jane heard a little quiver in her voice that she hoped he wouldn't notice.

Cheek leaned back, twitched his head and spoke softly in a creepy way, "You're right Ms. Koss. I did have something to discuss with you." He drew out a thick file folder and flopped it on the desk. "Our research on you. Go ahead and take a look."

Jane wanted to bolt out of the room. But she tried to look poised as she gazed at the thick file. *What could be in there? I lead a pretty boring life.* "Why should I look at it?"

"I want you to know that we understand that you do not share the same environmental views as your employer. On one of those pages is a list of causes you've contributed to and Facebook likes and comments that show your true feelings." Cheek smiled showing too much teeth.

Jane felt angry and betrayed. *How could they get this? It's none of their business.* "That may be so, but I'm working for Colonel Bolton now and none of that matters."

"Oh but it does matter. Because you are going to want to work for us." Cheek leaned forward.

"You are offering me a job? You want me to quit my job as administrator of the Colonel's 501?" Jane wanted to know.

"I am offering you a job, but we want you to continue working—at least outwardly, for the Colonel." Cheek's head jerked again.

"Wait. That would ruin my career if it ever got out. I would never administer another campaign. That's my life-long career goal," Jane reddened. "You would destroy me to help your boss. Well I'm not that way. I will not destroy anyone to keep my job."

"You would destroy the environment to maintain your position. What kind of job is that?" Cheek' voice started to rise again.

Jane tried to stand up abruptly, but the heavy chair did not move back, and she ended up just popping up and falling back into her lumpy chair. "I don't need to listen to this. I've worked hard to get where I am. You are not going to destroy what I've built." Jane wished her voice wasn't so high.

Cheek put a small iPad on the table. "I think you'd better look at this video before you make up your mind. We'd hate to leak this to the press."

Jane paled. *What could be in that folder? Sex photos? Only if they were fake! Illegal activities? No. She smoked some weed in college, but no one in Alaska would care about that.* "You have nothing on me. I'm clean. Whatever is in there is made-up." This time she was successful at standing up. Then she wondered, *How do I get out of here? Where are my purse and things? My phone and keys?*

Cheek just sat in his chair confidently staring up at her.

"Where are my things? I'm leaving." Jane crossed her arms in front of her.

"Sit down." He urged. "You'll get your things, but only after you watch this. Then you will want to stay much longer. Nothing in it is photoshopped." Cheek's yellow teeth flashed a mean grin as Jane sat down. He tapped the tablet's screen.

41

HOURS LATER, DAVE AND ETHAN burst into the computer lab. Dave held up Conrad's tablet computer like it was a trophy. Sully seemed to be the only one there, and he was on his knees probing under a desk with a broom handle.

"Where is everyone?" Dave asked.

"They're with Conrad trying to fix the software. We think a software glitch did a number on the data stream," said Sully as he manipulated the broom handle under the desk.

"What are you doing in Zhang's cubical?" Ethan asked.

"I'm getting Zhang's batteries. I think they rolled under this desk when I dropped his mouse yesterday. My batteries are dead," said Sully. "Has anyone seen Zhang?"

"I think Ethan and I found the reasons for the inconsistent data. Let's go to Conrad's office for a staff meeting." Dave held up the tablet as if that would explain everything.

"OK. I've almost got these out, can you a wait just a few minutes?" Sully asked.

"Sully, this can't wait," said Ethan. "We're going to share our findings with the group. This is urgent."

"Maybe there will be refreshments," said Dave winking at Ethan.

"Yeah. Maybe some donuts came in on that last supply helicopter," said Ethan.

Sully brought up his head bumping on the underside to Zhang's computer display making it rock dangerously, "You guys are making this up. No helicopter can fly in this storm. But I'll go—just in case."

ONCE IN CONRAD'S OFFICE, Dave got everyone's attention, "First, let's find out what you guys learned from Conrad's examination of the software."

Conrad stood next to his stool hardly taller than when he was sitting on it, "First of all, there are some efficiency problems caused by system resource errors. This really slows down the data transfer. It might just be sloppy programming or someone is intentionally slowing down the software."

"Does it look like we've been hacked?" asked Sully picking up Conrad's mouse.

"Wait until I finish," said Conrad taking his mouse out of Sully's hand and putting it in a drawer.

"Second, the readability of this code is really low. Programmers make code hard to read when they are in a rush or they are trying to do something sneaky. It's going to take me several days to figure out what all these programming elements are actually doing.

"And third, the algorithmic complexity is out of whack. As a result, the execution time for this software and the expressed resource use is way off."

"So what does that mean," Taylor wanted to know.

"It means we don't know if there's been any sabotage," said Dave. "Let me tell you what Ethan and I learned. We dug down into the uncombusted waste from that last burn and guess what we found?"

No one guessed, so Dave carried on. "We found mercury, uranium, thorium, sulfur and a high moisture content."

"High moisture content?" echoed Sully.

"What does that mean?" asked Taylor.

"It means you burned the wrong coal," said Sully. "Someone put some crappy brown coal in the combustion chamber."

"How could that happen?" asked Dylan standing off to the side.

"Last summer I labeled all the coal deposits for testing and someone screwed it up," said Sully. "Maybe a lazy tech just grabbed any old coal out of that mine.

"What do we do now?" asked Taylor.

"We go out and bring in some of the good stuff," said Sully. "It's our only choice. There's no more grant money until we demonstrate performance."

"Could burning brown coal screw up some of these data?" asked Taylor.

Conrad looked skeptical. "I'm not sure. Let's do another burn and see what happens. Meanwhile, I'll clean up this software."

Dave looked at Dylan, "You are the Alaskan outdoorsman. Can you lead the away party to collect a few kilos of coal? It's pretty stormy out there."

Dylan smiled. This was something he could do better than anyone else at ARCS. "Sure. I'll make a list of what I need."

"If you put donuts on the list, I'll go with you," said Sully.

42

ETHAN HEAVED THE FIRST huge duffle onto the lab table, unzipped it and dumped out the contents, then repeated it with the second duffle. "Here's the stuff you wanted, as much of it as I could find around ARCS."

Dylan started to sort through the big pile.

Sully scowled at the mound. "We need to take all this shit? We're just going to get a couple of kilos of coal. Whose idea was to bring so much stuff? I'm not carrying it."

Dylan looked at him. "I'm a professional guide. I do this all the time. More important than the coal, is that we remain safe out there." He gestured to the window where snow could be seen blasting horizontally through the bitter storm.

Sully picked up a roll of duct tape. "So why the tape? You're bringing a bunch of shit we don't need."

Dylan took the tape from Sully and put back into the piles he was making. "Duct tape can be used for blisters, clothing repairs, emergency wound dressings and it burns like crazy. It can serve as a fire starter. Now let me sort everything so the loads are distributed appropriately."

Taylor walked in carrying an armload of backpacks. "Sully, you will carry your share, or I'll assign you to bring back all the coal as well. Did you get the snow-shoes as I asked?"

Swerving to avoid a bearded technician, Sully slunk out of the room. "I'm not going to carry a pick and shovel."

Ethan watched him leave. "We're not bringing mining tools. There's plenty at the mining site. At least there was last summer when we were there."

In the back of the room, a bearded electrician with long red hair quietly worked on a nest of wires with a testing device.

SEVERAL HOURS LATER Sully, Ethan, Taylor and Dylan stood before the packs.

Dylan held up a pack. "We're going about six miles each way to the coal mine and back. We'll spend the night in the shelter outside the mine. To reduce the amount of gear we need to take, I've done away with redundancies except for the emergency response beacons. Each of us carries one of these ERBs."

"Which is the lightest pack?" Sully looked at the four packs lined up.

"Taylor has the smallest body and the lightest pack. Her pack is designed for a woman's body proportions. You wouldn't like it." Dylan indicated a red backpack. "This one is yours. It has your insulin, water, headlamp, emergency shelter and energy gels. You are carrying our first aid kit. You will carry your own MREs."

"MRE? What's that?" Sully peered into the pack.

"Meals Ready to Eat." explained Ethan as he explored his own pack. "Hey I get the sat phone, radios and stove kit."

After digging in his pack, Sully held up a box, "Hey anyone want to trade? This one is beef patty."

Dylan tossed him a box, "I'll trade it for jambalaya."

Sully looked greedy, "Deal. Buddy you got screwed on that trade."

Dylan looked away. He considered the jambalaya inedible, but didn't have the heart to give it to someone else. Now he didn't need to feel so guilty about letting someone have a terrible meal.

"Hey, who's going to carry me on this trip?" Conrad shuffled in with his ever-present coffee cup.

Dylan realized Conrad would want to go with his team, but would never be able to keep up due to his physical limitations.

Ethan looked at Dylan, "We pulled him out to the combustion station on a sled when we were setting it up." Then turning to Conrad, "This is an all-day hike each way, and we're carrying coal back here. You'll need to stay here and keep the coffee pot going."

Conrad smiled good-naturedly. "Sure thing. I'm an expert at that."

Ethan looked at Dylan. "He's our computer whiz. We'd be lost without him."

Conrad beamed.

Taylor pulled a red metal bottle from her pack. "Why are we taking bear spray? Aren't they all asleep in this weather?"

"That would be nice, but this is a shoulder season. Some bears become active and start foraging this time of the year. Better to be prepared. You should clip that to the outside of your parka so you can get to it fast."

Dylan walked over to Taylor and took the bear spray. "Have you ever used this before?"

Taylor cast her big eyes at Dylan. "No, how does it work?"

He handed the spray back and stood behind her. "See that white safety clip? Just pull it out and point. Aim for the chest of the bear because it will duck right into the spray. After you spray, be sure to move to the side. The bear won't be able to see and if it continues to charge, it will aim for where you were last seen."

She leaned back until she could feel Dylan's warmth, "Like this?" She pretended to spray Sully.

"Very funny. Shouldn't we take a gun? Don't you know anything about bears? They are dangerous." Sully looked nervous.

An electrician working in a power closet suddenly stiffened and stopped testing wires.

"If you want to pack a gun, go ahead, but I'm not going to carry it. Also, you'd better be experienced with firearms. I don't have time to train you."

To forestall further debate, Dylan didn't bother to tell Sully that the spray was much more effective against aggressive bears than firearms. He turned to his pack and checked once more where everything was packed.

Taylor felt her face had turned red, but ignored it as she realized she had enjoyed being close to Dylan.

"Everyone, let's meet back here after breakfast tomorrow, suited up for a trek in the snow. You should bring your goggles and full face covers." She didn't notice the longhaired electrician dash out of the room.

43

NEARLY INSTANTLY AFTER JANE tapped the screen on the small tablet, music flooded the room from hidden speakers, a slideshow of stunning Alaska wilderness and happy residents paraded across the screen in perfect time to the music. The music and photos had an unmistakably calming effect on Jane. She loved Alaska with all its quirky culture from the charming winter *Rendezvous* events in Anchorage, to the hunting celebrations of the natives in the far north.

Soon the music changed and a deep, rich voice that Jane had heard hundreds of times on voice-overs for movie trailers began speaking. "In a world where climate change is moving faster than anywhere else on earth. Where the melting glaciers, rising sea levels, flooding of coastal communities . . ."

Jane watched scenes of coastal destruction and displacement, dead and dying wild animals, lost-looking polar bears, and unhappy natives moving out of their traditional flood-damaged homes.

"Loss of sea ice cover and habitat for arctic species, thawing permafrost and increased storm severity . . ."

Jane watched a relevant and emotionally moving video and stills illustrating the voice over. She found herself becoming alarmed and sad.

"Infrastructure damage, changing wildlife migration patterns, loss of safe drinking water for people and animals, significant coastal and river erosion, more dangerous hunting and fishing for people and animals, pure rivers filling with silt, tundra displacement, invasive species surviving in a warmer climate . . ."

The photos and videos were becoming increasingly disturbing. Jane felt uncomfortable knowing what the ending was going to be, Reelect Governor Leola Bates-Hardy, for Alaska's sake. But the ending of the video was more powerful than she expected.

THE RICH VOICE SPOKE, " . . . so Jane Koss. You want Greg Yeung to be Alaska's governor. Greg Yeung, who says that job-creation trumps the environment. Greg Yeung, who thinks any environmental regulation is an impediment to progress. Jane Koss, this is your chance to do something important in your life. Join Governor Bates-Hardy's team and save Alaska."

The video ended with a smiling native child wearing a *Reelect the Governor* button.

Jane was stunned. The production costs on the video must have run into the many thousands of dollars. Hiring the biggest voice-over artist in Hollywood would be at least three times her monthly salary. All this was done to influence her. She was amazed at the expense and effort Cheek had put into this meeting.

"That's an impressive video, but it misrepresents Greg Yeung's stance on environmental issues," said Jane.

"Right. He believes market forces will moderate environmental damage," said Cheek, making air quotes around *market forces*. "Once there are no more salmon, then people will stop overfishing. He's crazy. You must know that."

"If you brought me here to try and convince me to change sides, it won't work. Besides, I don't set policy; I just administer the 501(c)(4). Now I need to go."

"You need to look at your tablet first," smiled Cheek with a flash of yellow.

Jane looked down at her tablet to see documents that had been on her office computer just that morning. A quick glance showed Jane that Cheek knew about the cash flow problems the 501(c)(4) was having. This made her mad. *How could he get these?*

"You hacked my computer. That's illegal. Now you are committing criminal acts. How's jail sound to you?" Jane tried to sound confident despite being unnerved.

"And just how do you intend to prove we hacked your computer? Are you going to take that tablet with you? I can wipe the hard drive remotely before you are out of the building. No, you won't bring charges, but you will join our side. Bolton's 501 has overspent, and he's desperate for money. You don't want to tie yourself to a sinking ship."

"So you think I want to join a law-breaker's side and become a criminal?" Jane asked.

"The governor gives me a long leash. She knows nothing about the hack. You wouldn't need to work in my department. You could get a position on the winning side. You could work for the Bates-Hardy 501(c)(4)."

"No thanks. Even if Greg Yeung doesn't win, I can get a job with another, bigger 501 based on how well I run this one. Now it's time for me to go." Jane stood up.

"There's another reason you'll want to join our side," grinned Cheek.

"And what's that?" snapped Jane.

"To stay out of jail."

44

IN THE BUNKROOM, Dylan spread the clothes from the duffle that Anja had packed for him. It looked like she had gone to a sporting goods store and just threw things into a cart. The base layer was all cotton, so it was worthless.

He ignored the expensive new boots. Only a rookie would start a hike in new boots. The socks looked pretty good. The snowsuit was excellent except that it might have been designed for a teenager, with wild colorful geometric patterns that would give him the appearance of trying to look hip. Dylan figured he could wear the polyester pajamas under the suit as a base layer. He wondered why Anja would pick out cotton base layers and polyester pajamas.

AFTER BREAKFAST but before dawn, the away party suited up, donned their packs and gear and stepped out into a rosy crystal wonderland. Sometime during the night, the storm abated and left the world awash in luminous dawn-pink snow-frosted brilliance.

As Dylan stepped outside, he realized he'd recently been in a building or plane for far too many hours. Energy poured into him from the tiny ice crystals suspended in the cold, clear air.

Dylan's mind cleared as he felt himself absorb the elusive view of the massive Denali summit, smells of the snow-laden trees and the sounds of a chilly breeze through frozen tree branches. His joy was tempered by the knowledge that the weather would turn by day's end.

After about 30 minutes of hiking, the party had to stop to remove some clothing. Everyone was hot.

"Why don't we go back and get the snow cat?" Sully threw down a Snickers wrapper and pulled his jacket and then a fleece layer off. He shoved the fleece into his backpack and dug around for something else to eat. "We can ride in style."

"Hey Sully, does Bambi shit in your bedroom?" Dylan looked down at Sully's discarded candy wrapper.

"What the hell kind of question is that? No, of course not."

"Then don't throw your crap on the ground. If I'm on a hike, we *leave no trace.*" Dylan pointed to the wrapper.

Sully picked up the trash and shoved it in his pocket. "If we had the snow cat, we would be able to eat real food."

"I already told you that we'd need to go clear around this forest. It would take us an extra day in a snow cat. Besides, the weather is supposed to get worse by noon and be back to near-blizzard conditions this evening. A cat could get trapped by wind-driven snow banks." Dylan put on his lighter gloves.

Ethan tossed a snowball at Sully. "Hey, cheer up. Maybe we'll need to be rescued, and you can have a snow cat ride back." Dylan had started to feel a vague sense of dread about this outing, and he knew that Ethan did too.

After another 30 minutes of hiking, the party decided to rest near the combustion station.

Dylan slapped the metal side of the brown metal structure, "What's this doing out here?" A small metal access door flopped open at the impact revealing a panel of switches, gages and data ports.

"It's where samples were burned remotely. It also collects and stores raw, unprocessed data from our labs. We built this far from ARCS in case of an explosion." Taylor explained. "It's only a 20 minute walk from ARCS in dry, calm weather, but it's an hour by snow-shoe and dragging Sully."

"Hey, I have foot issues." Sully said defensively.

"Why did someone leave a boot out here?" Ethan pointed to a boot stuck in a nearby snow bank.

Dylan felt uneasy when he saw the boot near a small mound suggesting another boot. He walked over to examine, and nudged the toe. It wouldn't budge. He bent down to pick up the boot, but he couldn't. It was attached to a frozen leg.

IN THE TREES, a longhaired man in snow-camo put down his binoculars. He toggled his secure walkie-talkie. "They found him. Should I shoot them all now?"

45

AFTER DIGGING FOR a few moments, Dylan and Ethan were able to expose a body that had been buried in the snow bank. Taylor pulled back the stiff, frozen hood to reveal a face.

"Oh my god! It's Zhang!" Sully started to bellow. "Oh my God! Let's go back. I don't like it out here."

"He's frozen, but I wonder if that's the cause of death." Taylor examined the corpse for obvious signs of trauma. "I have never seen a body with such a look of terror on the face! It's like his face froze at the moment of death."

Dylan dug around in Ethan's pack for the satellite phone. "We need to call ARCS so they can get a sled out here to transport this body."

"Let's just go back. God, I'll never be able to forget that face! You can get the sled at ARCS and come back for the body." Sully backed away from Zhang.

"I don't think we should go back. We need the coal sample ASAP and going back will do nothing to help Zhang." Taylor stood up. "I can not find any signs of trauma."

Ethan pulled Zhang's pack out of the snow and shook it off. "Let's see what he was doing out here." Ethan managed to get the zipper open and dumped out the contents of the pack.

"Hey, that's my Hytech Ms.2921a!" Sully overcame his discomfort and grabbed the testing tool out of Ethan's hands. "See, here's my initials and here's a smear of peanut butter on the side. He never asked me to borrow it."

"A portable nuclear magnetic resonance spectrometer, cool! The last time I saw one of those I was in college and they were twenty times bigger. I used it in the lab to analyze chemical reactions," said Taylor.

"It looks heavy," said Dylan as he opened up the satphone and prepared to make a call.

"Correct. Heavy. It has magnets, but it's really handy to study multivariate correlations, but this model can do all kinds of data analysis. We can use it to estimate the number of alcohol molecules created by *the Scrubber*. It's basically a portable testing device," explained Ethan.

"It can also function as a spectrophotometer. We can use it to count bacteria in the spent media. And it's mine. All of my readings are stored on it. I paid extra to get it loaded with memory." Sully cleaned snow off the device and scratched at the peanut butter smear with a gloved finger.

"Maybe it will tell us what Zhang was doing," said Ethan.

Dylan punched the keys on the phone. "This isn't getting a signal." He turned it over. "Criminy, who bought a WorldSat Phone?"

"I did. So what?" Sully looked up from packing the spectrometer into Dylan's pack.

"*So what?* WorldSat does not have low earth orbit satellites in the north. These things are designed to work for southern hemisphere needs. Why are you putting that heavy testing thing in my pack?" Dylan sounded angry. "If you want it, carry it yourself."

"We might catch a sat-phone signal at night. Let's try calling ARCS tonight." Ethan took the phone from Dylan and put it back into his pack. "Meanwhile, let's light up an ERB on Zhang's body so a party can find it tomorrow. As soon as the ERB is turned on, Dave will get the signal and send someone out here."

"Use my beacon. I need the space." Sully pulled the candy-bar sized device from his pack, turned it on and dropped it onto Zhang's body.

Dylan bent down and tucked the ERB into Zhang's pocket. While low in the snow, he saw a flash from some trees about 1,000 yards away. It reminded him of the kind of sun flash that comes off a riflescope. He shook his head and thought, *It can't be. Who could possibly be out here hunting? That flash must have come off some ice.*

After a moment of silence around Zhang's body, the group continued their journey in a solemn mood.

SEVERAL HOURS OF SLOW hiking later, Taylor stopped and pointed. "We've been here for five weeks, and this is the first time I've seen Denali's summit."

Taylor gestured toward the massive mountain. "It's probably the most amazing sight I've ever beheld."

Even Sully seemed moved by the view of the mountains. "You can thank the Hines Creek Fault for this part of the view. It's allowing Denali to grow by over a millimeter each year."

Ethan shaded his goggles and looked at the snowy brilliance of Denali. "That's not much."

"Oh yeah? In two million years, an eye blink in geologic time, that means a gain of a kilometer. Of course, by then they'll be no snow on it. The carbon dioxide in the atmosphere will warm up this place plenty. Since Denali is mostly granite, it doesn't erode fast so we'll have a tall, bare mountain here." Sully gazed at the peak as if he could see it grow.

Since they were stopped, Dylan studied a bank of dark clouds rising in the north. "We need to keep moving, if we are going to sleep in the shelter and not in a snow cave." His voice was calm, but the others picked up on some urgency.

Something besides the storm was bothering Dylan; something didn't feel right. He couldn't say what it was, but he noticed he was feeling increasingly uneasy since they left Zhang's body.

Soon a drop in temperature put a crust on the snow that made it much easier to hike, but forced a halt to allow everyone to upgrade to warmer gloves and close snowsuit vents.

Dylan took the time to talk with Taylor. "We need to pick up the pace. That front is moving much faster than the weather reports suggested. We are not prepared for a blizzard away from shelter."

Taylor nodded. "Hey Sully, need some help getting that fleece back on? We got to move out." Sully was having trouble getting his arms into his fleece. He was obviously cold with his jacket off and it was quickly filling with wind-driven snow.

"My feet hurt, and I'm cold." Sully's teeth chattered as he allowed Taylor to help him with his gear.

Dylan looked concerned. "Your feet hurt? What's the problem?"

"I feel like they are freezing and all torn up. It's weird because these are brand new boots and they are the right size and everything."

Not wanting to carry Sully, Dylan hunched over Sully's boots and took one off. He was wearing three pairs of cotton socks, now bloody at the heel. "Why don't you have on the wool socks I put in your gear pile yesterday?"

"Wool? Are you shitting me? They itch." Sully grimaced as Dylan pulled off the last sock. His toes steamed in the cold air.

"You should say something the moment you feel a blister starting so we can tape it up." Dylan cut off the cotton socks. Soon Dylan had Sully's wounds treated and Ethan's spare socks on him. Dylan used the duct tape to pad Sully's blisters.

"Anyone else feeling a blister coming on? We should tape it now so it doesn't get worse." Dylan held up the tape. No one showed interest.

A blast of wind hit the party from behind sending a chill up Dylan's spine. "We got to get moving. Our visibility will drop to zero if that storm is as bad as it looks."

From the woods below, another flash of light caught Dylan's attention. *Weird. No one is out here except us.*

46

JANE FELT ANNOYED that her hands shook as she gazed at the tablet's screen. She watched a slideshow that had images of what appeared to be copies of financial documents related to Colonel Bolton's company. Some appeared to be internal memos and one caught her eye.

It was a bad photo of a message from her to the Colonel telling him she knew about the stock fraud he was perpetrating on his investors.

Jane suddenly felt cold.

"Where's the rest of this memo? I told the Colonel to restore the money, or I'd quit."

Cheek stood up. "I must have forgotten to include that part. It looks like you were privy to Bolton's stock fraud. You are likely an accessory after the fact. I think that's probably three to eight years in prison."

"You're threatening me." Jane found herself close to tears and angry with herself for feeling that way.

"Think of it as we want to help you avoid prison. All you need to do is bring us Bolton's confidential files by this time tomorrow. You notice we can access your files, but not Bolton's. If you help us, we'll forget your part in this little stock fraud."

"Fuck you," she said.

"If you don't want to help us to save yourself, how about doing it for Alaska?"

"I'm working to get Greg Yeung elected for Alaska. Then assholes like you will be out of a job." Jane thought she sounded more confident that time.

"Yeung will harm Alaska. His plans include removing nearly all regulations on oil exploration and drilling. Imagine massive oilrigs off the remote and pristine coasts of Alaska, the Beaufort Sea, Native lands and even state preserves. All the inevitable oil spills will destroy some of these places." Cheek's stale breath wafted over the huge desk at her.

"That's not part of Greg Yeung's plan for Alaska. He's in favor of opening the Arctic National Wildlife Preserve for exploration, but that's it." Jane felt certain of that.

"If you worked for me, I could show you the ropes. We could become very close," Cheek flashed a smile Jane figured was his attempt at looking sexy.

"Where are my things?" Jane crossed the room holding the tablet and opened the door to the anteroom, "Mr. Small?" she yelled. "Mr. Small, I'm ready to go now. Please get my things."

She glanced over her shoulder to see Cheek behind his big desk. She wanted to bolt from the room, but wasn't sure which way to go.

"Jane, all you need to do is say you are not interested. There's no pressure. You can keep that tablet. Let me know if you want the originals," Cheek tried to sound reasonable. "If you come over to our side, you won't be dealing with me at all. All your communication will be with my assistant, Mrs. Blodgett."

Jane yelled again, "Mr. Small? Please bring my things."

"Ms. Koss, we don't want you and the Colonel going to jail, you'll be helping the environment and you'll be well paid working for us." Cheek sounded confident.

Unable to remain in the same room with Cheek any longer, Jane burst out of the huge office and turned left at the first hall. Running down the hall in the slippers given at the security screening, she came across three workmen in hard hats apparently changing a light bulb.

Breathless she said, "Which way to the security screening room?"

Two of them pointed back the way she came and the other pointed the opposite way. "You can cut through the big kitchen and get there pretty fast. Just keep to the west."

Jane slowed to a walk and proceeded through the big, modern, kitchen. At every junction, she turned west and soon found herself in the security room with two troopers who had nothing to do.

"I'm Jane Koss," she held up the name badge she had been given. "I need my things, purse, phone, keys and shoes."

The female trooper stood up. "Of course Ms. Koss. Wait here and I'll get them for you."

Twenty minutes later, Jane was in her little X90 and rumbling down the gravel drive. She was shaking and crying. I don't want to go to prison. I've worked so hard to become a Super PAC administrator. I don't want to give up my dream. I don't want to harm the environment. Does anyone else know what a horrible person Harold Cheek is?

As she poured through the rush of thoughts cascading through her mind, there was one thing she knew for sure; she had to talk to the Colonel right away. He would know what to do.

47

A BLAST OF WIND hit the party from behind momentarily stopping Sully's constant complaining. This time the wind gusts bore ice particles and snow. Dylan knew the direct route to the coal mine lay to the west, but he decided to take the party on a slightly longer route to take advantage of some cover that the forest to the east would offer.

Their snowshoes were hanging on their packs but each member of the party wore traction cleats over their boots. They were starting to make good time.

Dylan thought again about leaving the emergency beacon on Zhang's body. If the staff at ARCS sent out a search party in this bad weather, someone else could get hurt. Then Dylan wondered how Zhang died. It didn't make sense that such an able-bodied man would just die of exposure when he was wearing all the right gear.

As if reading his mind, Taylor caught up to him. "So what do you think killed Zhang? Have you seen anything like that before?"

"I've had clients die on a trip."

Taylor watched as Dylan's attention seemed to leave, and he stood still. Taylor thought he looked like he was having a PTSD episode.

Dylan roused himself, "I've seen people die from exposure, falls, gunshot or other trauma." Dylan thought of a time long ago when he was nearly killed by a group of mercenaries who were after him. He suppressed a surge of anxiety just thinking about what he had done back then. "But I have no idea about the cause of Zhang's death."

Taylor nodded.

"What about Sully?" Dylan asked. "Is he in bad shape or is this the way he always is?" He looked over at Sully who was trying to get Ethan to carry some of his load.

"He's always like this. I know he can be annoying, but he's really an amazing scientist. He was Dave White's first pick as a geologist for this project," Taylor said. "I think we should take a break once we get into the woods so I can check his feet again. He has circulation problems, and he might not realize how bad he is."

"I'm concerned about this whole project," Dylan confessed. "I don't think my CO_2 scrubber is a great idea. I'm wondering if it's being used because of politics and not science. Right now there are probably hundreds of devices that metabolize CO_2. I don't think mine is especially great."

"Professor White looked at scores of designs. Yours fit all his parameters. Why do you think Professor White launched this project using your design?" Taylor asked.

"I've been asking myself that. My daughter controls Mountain Club political money and an army of campaign volunteers. Governor Bates-Hardy needs those resources to counter a political challenger from the right. I can't stop the feeling that I'm a cog in a huge machine. Maybe Governor Bates-Hardy came up with the grant for this project just to get reelected."

"If you think that, why did you agree to come all the way out here?"

"I mentioned Anja the other day. I'm trying to form a relationship with my daughter. I didn't even know she existed until recently, and I'm trying to find a way to be part of her life." Dylan suddenly felt better talking about his worries. He wished his wife, Suzie, were with him. She would be able to soothe his concerns and offer solid advice.

"Is it time for a rest yet?" Sully shouted from the back of the line. "I'm ready for a break."

"We're almost to those trees. We'll take a 10-minute break where the wind should be milder," Dylan called back. He turned to talk to Taylor but found she was walking next to Ethan and deep in conversation about carbon dioxide metabolization.

Once the party entered a relatively quiet wooded area, Dylan declared a ten-minute break. He said it would be their last break before pushing on to the mine and the comforts of a miner's shelter. Dylan found a protected area near some blown-down trees. Sully immediately sat on a log, pushed out his feet for Taylor to attend to, and started playing with his nuclear magnetic resonance spectrometer. Pulling off Sully's boot, Taylor removed most of the tape on his foot and looked for problem areas.

"Hey! What the fuck?" Sully yelled.

"Did I hurt you?" Taylor stopped the process of applying fresh, dry gauze onto his blisters.

"Ah, no. I mean, look at this! Hey Ethan, come here. Look at these data. Something's really wrong." Sully wiggled his toes unconsciously.

"What's up?" Ethan pushed himself up, sat next to Sully and peered at the screen of the recorder. "What am I looking at?" Ethan asked.

"You're looking at about ten times the energy we expected going into *the Scrubber*. This is impossible."

"Whoa. Are you sure you are looking at the right data? I studied all the power consumption data at ARCS. There was nothing like this in any of the read-outs." Ethan frowned at the screen.

"Dylan, we need to go back to the combustion station. Zhang must have had a power surge or something that corrupted these data. If this information is correct something is seriously wrong with the entire ARCS project." Sully stood up putting his feet into a snow bank.

"Sit down. I'm not done." Taylor ordered. "Don't get so excited. When we get to the shelter, we can all study the data Zhang downloaded. There's probably a reasonable explanation for what you found."

"You know," Dylan stared into the trees, "this energy spike is yet another indicator that *the Scrubber* is not working. There are three critical issues we need to explain. First, there were the high CO_2 levels."

Taylor climbed into Dylan's thought stream. "Then there is the fact that we initially–detected live bacteria downstream from *the Scrubber*, and then we couldn't."

"That's right, Taylor. Thirdly we detected toxic metals; maybe bad coal. What the hell is going on?"

Dylan shook his head and brought himself back from the cerebral world of science theory. "We're not going back today. It will be dark soon, and I don't want anyone to step into an unseen hole," said Dylan.

"Besides, we are nearly at the mine. Let's get the coal we need and then go back tomorrow," Taylor said.

FROM 400 YARDS away, a soldier in snow camo put down his parabolic mic and thumbed his walkie-talkie. "They know. What are your instructions?"

SULLY SQUINTED AT SEVERAL plastic-wrapped packets in his lap as a sharp wind blew through their resting area. "Hey, Ethan. You have my water. I need it to activate the flameless heater in this MRE.

"I don't have your water, I just have mine." Ethan closed his hood as a loud wind roared in the treetops above.

"No, I put my water in your pack during the last rest stop because I have so much more weight than you." Sully held up the data reader.

"You didn't ask me, so it's my water now." Ethan gave a wink to Dylan.

"Hey Dylan, can I have some of your water? I need it for this MRE."

Dylan suppressed a smile. "Sully, save your MRE for dinner. We need to move out as soon as Taylor is done with your feet. And drink your own water."

Sully looked down at Taylor. "Can I have some of your water, but just for a drink? I'll eat these crackers for now and save the MRE for later."

Taylor paused in her foot wrapping to get her water bottle and a big first aid kit. "You can have some of my water if you carry this first aid kit."

"But that wouldn't be fair!" Sully wailed. Seeing that Taylor and the others had formed some kind of alliance, Sully took the first aid kit, exaggerating the weight, and then the water bottle.

Dylan picked up the spectrometer and showed it to Sully. "Sully, show me where Zhang's data indicates that no bacteria are present in the spent media."

Sully's fingers stiffly moved over the keys and touchscreen. "It's right here. The data at ARCS shows spectrophotometric readings consistent with high levels of bacteria coming out of the membrane. That's what you would expect. The bacteria are happily growing in the membrane eating up the CO_2 and some of them are washed out of the membrane as the new growth media arrives."

Dylan nodded, "That makes sense."

Sully continued, "But the unedited data from my tablet shows different spectrophotometric readings—readings indicating no bacteria in the spent media coming from the membrane. No bacteria in the spent media means no living bacteria in the membrane, which means that our current generation scrubber is not removing CO_2 from the exhaust gases."

Sully looked up from where he was tying his shoes. "So we got to go back now. If the original data is correct, the governor's announcement of a breakthrough will be false."

Taylor saw where Sully was going with his argument. "If the governor makes a fool of herself, that anti-science nut Greg Yeung will get elected. It would suck for Alaska."

"While you are pondering those prospects, here's another interesting observation. There is a definite spike in power consumption, and I can't explain it! I need more time to study these readings," said Sully as he stared at the screen of his spectrometer.

Dylan felt let down. He never really imagined that his college experiment would save the world, but part of his mind did not want to abandon the idea. "Isn't it possible that Zhang did something wrong when he downloaded the raw data? Before we decide *the Scrubber* doesn't work, shouldn't we reexamine everything? Redo the experiment with better controls and so forth?"

The others just stared at Dylan as they slung their packs. Crumbs fell like snow from Sully's jacket.

"We can't go back to ARCS now anyway. We're nearly at the mine and this storm could get deadly." Dylan pointed out. "Maybe we'll have some sat-phone signal at the shelter, and we can have the ARCS team recheck the raw data when they get Zhang's body."

Just then Ethan jerked back, his throat erupting in a spray of blood and tissue. Droplets of blood and a pink mist sprayed over the entire area spotting everyone's clothing, and releasing a strong coppery odor. Ethan fell to the ground, as arterial spurts from his open neck quickly drained his quivering body.

Wondering at the unimaginable event that just happened, Sully and Taylor looked stunned and unable to move. Near Sully's seat, a small explosion of bark and wood chips told Dylan that rifle fire was coming in.

"Get down!" Dylan charged at the two knocking Taylor and Sully back over a log as the place where Sully was sitting puckered when a bullet slammed into it.

"Hold your fire!" Dylan yelled in the general direction of the shooter. "You're firing at people."

His call was answered by several more shots hitting nearby.

"What's happening?" wailed Sully.

"Holy shit! They are trying to shoot us. We got to get out of here." Dylan pointed to Sully's left, "Quick. Go that way and stay down."

Sully had tears in his eyes. "I'm not going without Ethan. He's hit."

"He's dead. There are bad guys out there shooting at us. They are killers! You can't help Ethan now." Taylor, stuck in a narrow space behind Sully and in front of Dylan, pushed Sully. "Go! Go! Go! God damn it!"

As the group moved out through a thick cover of downed trees, Dylan cast a quick look back to see who was shooting. Whoever it was, appeared invisible in the blowing snow.

Quickly the group scrambled along a line of trees trying to keep low. Dylan noticed the strong smell of Ethan's blood on his jacket, and his hands stung like fire as his naked fingers propelled him through the snowy woods. Silenced shots! The very thought brought a flashback that Dylan had to suppress.

"Go into the wind!" Dylan yelled up at Sully. The storm blew snow and ice pellets horizontally. It would be difficult to put a scope on someone if the shooter had to expose his face and scope to the biting wind. Sully followed the directions.

Dylan looked back and could barely make out three white-clad figures attempting to follow them. Running into the wind after an arduous hike, Dylan knew that Sully and Taylor would not last long. He had to come up with a plan. He stopped the pair near a high sparsely wooded place that would be a riverbank in the summer.

"I'm going to try and lead them away from you two," Dylan yelled over the wind. "You're going to need to trust me."

"What should we do?" asked Taylor.

"You're going to jump down this snow bank into the river channel. There will be no water in this channel until the spring melt. I'm going to throw these old dead branches into the hole you make when you jump to try and hide your position. Then I'm going to lead these guys on a path away from here."

"Where should we go after you lead them away?" Sully wanted to know.

"Just stay here. If the riverbank is eroded like I think it is, you'll be fine. There will be an air pocket where the bank overhangs the river bed." Dylan made a hurry-up gesture and Sully leaped down the bank. He landed up to his waist in snow, then disappeared completely as if falling into a hole.

Dylan made a motion for Taylor to jump. As soon as she was gone, Dylan threw part of a dead tree down the hole. It was a terrible job of hiding the hole, but Dylan hoped it would be good enough.

He started trampling snow and making a path away from the frozen riverbank. He didn't know if it was the wind or fear that made his eyes water as he ran from the faceless soldiers pursuing him.

49

AS JANE DROVE the rural highway away from the garish and vulgar governor's mansion, she felt an emptiness she had never felt before. What had just happened seemed so unreal that she wondered if she'd wake up on a beach somewhere and shake it off as a bad dream.

She knew her life was over now. All the work she had done to get herself where she was, now counted for nothing. She would be fired or forced to resign. Word would get around the small Super PAC communities, and she would never work again in her chosen field.

She might even go to jail. That would kill her parents. They were so proud of her accomplishments, and now this. She slammed her hand against her steering wheel. This would destroy the Colonel as well.

Cheek's actions had made it so she could never be in the same room with him. He was without any scruples, and she knew if it suited his goals, he would use the financial information to ruin her life and anyone else associated with the Colonel.

If she tried to charge him with anything, he could produce the documents to make her regret speaking up.

Jane's attention moved to the present. She had to call Bolton and warn him. There's no way she could let him be blindsided by this kind of attack by Cheek.

Jane pushed the big button on her hands free Bluetooth device clipped to her sunshade. Then she spoke, *Call Colonel Bolton.* The device answered, *Call Starbucks. Is that right?*

The damn thing never worked right. She fumbled for the off switch on it and pulled over, entered the number manually, switched on the hands free device and resumed her drive into town.

The Colonel's voicemail picked up. She left an urgent message for him to call her back. She then pulled over and called his secretary at her home.

"Elizabeth, I need to talk to the Colonel, but his voice mail keeps picking up."

"Of course it does. He and all his foreign fundraiser friends left all their phones in my desk and went inside the lounge area for a meeting. No one is to disturb them."

"Yes, but this urgent. I need to talk with him right away." Jane said.

"Well, you'll just need to wait. I would guess he'll be busy quite late. Why don't you try tomorrow morning?"

"Elizabeth, I'm just going to show up and demand to talk to him."

"Oh no you won't. There's even some foreign security people in the foyer. The elevator is locked down. I've never seen anything like it." Elizabeth sounded a bit awed.

"OK. I'll try him in the morning," Jane lied.

"That's a good plan. If you show up, you will just irritate a bunch of angry-looking people."

Jane ended the call. She had to try to see him. Once Cheek leaked the information to the press and the attorney general, the Colonel was sunk.

She wondered why it hadn't already been leaked. Maybe Cheek was trying to drive her to Bolton. Maybe he wanted to make sure they'd circle their wagons and protect each other. He certainly made it impossible for her to ever work on the governor's side of the fence.

What he had said about the environment had made sense. Certainly the Colonel's activities and political choices did nothing to slow the growth of carbon dioxide in the atmosphere.

Jane had never met the governor, but wondered why she would keep a person like Cheek on her re-election staff. It was said he was a political genius along the lines of Carl Rove. Rove had been able to get George W. Bush into the White House. If Cheek could get a governor turned pro-environment reelected in conservative Alaska, then he would truly be a genius.

She thought about the information Cheek had passed on to her. All of it caused short-term harm to the governor's campaign, and much of it pushed Greg Yeung into solidifying his position in favor of oil extraction, but against coal.

Being against coal would not likely harm a candidate in an Alaska election. The Alaskan coal deposits were of low quality and insignificant to most voters. It made the governor seem warm to the idea of encouraging coal extraction. Why would Cheek want the governor to appear pro-coal?

She thought about Colonel Bolton's fraud. He had taken millions from his investors and put the money at risk. At one point, he had to sell other investments at a loss to cover his deceit, but a rising stock market had made all the money problems go away. He swore he'd never do it again after Jane had discovered his actions and threatened to quit. *Can I trust him?*

He was not a great boss, but he was probably more honest than many other industrialists. He deserved to know about the files on the tablet sitting on the seat next to Jane.

She pulled into the hotel-parking garage. She was about to find out what would happen if she interrupted his big secret meeting.

50

THE STORM BLOWING above them, Sully and Taylor crawled under an overhanging riverbank toward a frozen mud wall. The air pocket under the snowy landscape allowed Taylor to nearly stand up straight under a frozen overhang projecting out above the frozen riverbed by nearly two feet. They were out of the wind and feeling warm for the first time in several hours. Dylan had picked the perfect place to jump.

Instead of relief, Sully wept, his shoulders shaking. "I can't believe Ethan is gone. He's gone. Just like that. It doesn't make any sense. Why are there killers after us?"

"Sully, I know how you feel, but we need to pay attention to our own survival and mourn Ethan later." Taylor noticed that the overhanging bank above their heads looked unstable despite being frozen. "Let's use these limbs Dylan threw in after us to prop up this river bank. It looks like it could collapse on us.

Just as she spoke, a mattress-sized slab of riverbank fell just downstream from them. The pair fell to propping up a much larger slab just above their heads. Even with makeshift support beam, they wondered if a roof collapse would bury them.

"Hey look at this!" Sully yelled as he knelt on the flat stone floor of the dry riverbed. "You know what this is?"

Taylor looked down at the bumpy riverbed. "Sully, don't yell. What if those guys are within ear shot?"

"These are dinosaur footprints! I've heard about the fossils in this area! It looks like a predator/prey drama is preserved here." Sully whispered loudly. "We got to come back here later and photograph these." Sully attempted to follow the prints as they disappeared under a dark, 12-inch gap between the frozen riverbed and the snowpack above. The prints made a path that vanished several yards into darkness. It was obvious Sully was too big, especially wearing his backpack, to explore the fossilized record of two parallel sets of footprints.

"Jeez Sully, you are so damned impulsive. One minute you are crying, the next you are excited about some depressions in the mud."

A loud *crunch* from the downriver side of the riverbank got their attention. Another crunching noise and the sound of scrambling came from just a few feet from Sully.

"Dylan?" Sully called. "We're just upriver from you."

Instead of the gaudy snowsuit that Dylan wore, a pair of legs clad in snow-camo crashed into their riverbank shelter.

AS HE RAN back in the direction of ARCS, Dylan stole a look back at his pursuers. There were only two that he could make out. He hoped the third was not looking in the hole where Sully and Taylor had jumped. Seeing a quarter-mile gap between prey and predator, Dylan changed his pace from a run to a jog as he drew his pursuers away from the dry riverbed. As exhaustion crept into his body, he hoped the guys chasing him were also feeling it.

At last Dylan came to the faint tracks left by his party earlier on their trek from ARCS to their last rest area. He noticed a candy wrapper, undoubtedly left by Sully.

Dylan ran in these tracks towards ARCS for a few yards until they petered out in an exposed area. Then he retraced his steps and followed the tracks going in the direction of the rest area. His goal was to leave a false trail that would force the team behind him to split up.

Once at the rest area, he tried not to look at Ethan's quickly freezing body as he dug through the dead man's pack looking for the satellite phone. It was quickly apparent that Ethan's pack had been plundered. Dylan pulled the gloves off of Ethan's hands and put them on. At least now he wouldn't need to run with his hands in his pockets. He'd lost his gloves somewhere when the shooting started.

Feeling he had no time to waste, Dylan walked backward in the steps the snow-camo guys had made when they ran into the rest area. Quickly he found the place where the men had observed the rest break. Dylan followed their tracks to an observation site used by the shooters.

Apparently, the team of killers used a machine unlike any he had seen before. It was about the size of a small park bench but shaped like the track on a military tank. Behind it were two open sleds big enough for a four-man team and gear.

Dylan, an accomplished handyman, understood small engines well, but this device appeared to be all-electric. In the woods, this would be quiet, unlike noisy gas-powered snowmobiles. Opening a side panel, Dylan saw a battery array that looked like it was taken out of a Tesla. The keys were still in it, so Dylan threw the camping equipment, surveillance gear, and electronics into the sleds and took off. Too bad they had not left a weapon behind. Dylan found a horn button and gave a long shrill blast before engaging the throttle. He wanted those snow-camo guys to chase him.

The electric snowmobile was fast, and after about a mile, Dylan found what he was looking for. He punched the throttle to the max and headed for a steep drop off. Rolling off the sled just before the heavy machine launched itself over the edge, he watched the contraption slam against some car-sized boulders and break apart hundreds of feet down a steep canyon. A miniature avalanche followed, burying nearly all trace of the device and its cargo.

Comfortable with the distance between himself and his pursuers, Dylan began a slow, cold march to where he left Sully and Taylor. He left no footprints on the icy ravine edge as he trudged along hoping that Sully and Taylor were safe. Just then he heard a muffled explosion in the direction he was heading.

51

PARALYZED WITH FEAR, Sully and Taylor watched as the snow-camo-clad legs dangled six feet above the frozen, dry riverbed. The crusty snow in the top layers prevented the broad-shouldered soldier from getting all way down to the riverbed. Then with a crash and a loud *snap*, the soldier, his pack and the biggest, longest rifle Sully had ever seen dropped onto the dinosaur prints.

The soldier seemed dazed by the fall and somehow Sully got himself moving. He grabbed the rifle by its long silencer and flung it spinning into the black space under the dark ice covering the frozen riverbed.

Sully and Taylor then started to scoot away down the tunnel under the snow formed by the river overhang. They managed to scoot to the opposite side of the tree branches, which had been serving as support beams for the sagging overhead riverbank. An ice wall halted further progress. This made for a pitiful hiding place, but it was about all they could do.

They watched as the camo-guy calmly opened his coat and pulled out a chrome pistol, pointed it at them, then toggled a shoulder mic.

"Concord to Tango."

Sully, speaking through tears addressed the soldier known as Concord, "Listen. You don't need to do this. We were just out for a hike. We didn't actually see who killed Ethan. You can let us go. We won't tell."

Ignoring Sully, Concord again turned his head toward his mic, "Concord to Tango, come in."

"Honest. We can pay you. I brought my own check-book," Sully blubbered.

"Tango, if you can read me, I'm at the river bank. I've located two of the targets, but I'm injured. I think I broke my leg."

The soldier looked up at Sully and Taylor. "Where's the data recorder? I need to destroy it."

"Hey," said Sully. "You are the state trooper from ARCS! Did you know someone dressed just like you shot Ethan?"

"Shut up. I need the data recorder," Concord said.

"No way. I paid nearly four thousand dollars for it. You can't have it." Sully had stopped crying.

"Sully! Give it to him. For Christ's sake! He's going to kill us." Taylor dug around in her pack.

"Tango, the data recorder is here. I don't have visual yet, but I can take care of these targets, and then search their packs. I hope you can read me, because I'm not getting any signal from you." Without emotion, Concord pointed his pistol at Sully. "You get to be first."

Just then an amazingly powerful red stream of bear spray hit Concord in the chin. He instantly ducked down, and the spray bore into his eyes, the powerful stream blasting between and under his eyelids. It filled the entire area with a chemically burning pepper cloud. Screaming, Concord dropped his gun and brought his hands up to his eyes, "My eyes! My eyes!"

Sully crawled over to him coughing, grabbed the pistol and threw it spinning into the darkness of the riverbed. "Taylor, let's get out of here. He can't see, and his leg is broken."

Taylor, also half blind from the bear spray, started edging farther away from Concord and the irritating pepper cloud, when a voice stopped them. "Halt. I can't see you, but I can stop you." In his hand, Concord held a grenade. Blindly, he fumbled for the pin.

"Wait, if you throw that in his confined area, the concussion and shrapnel will kill all of us." Sully continued to inch away from the blinded soldier.

"My death in support of our cause? It's a small price to pay." He pulled the pin and tossed the grenade in the direction of Sully and Taylor. It bounced off one of the branches holding up the riverbank and back onto Concord's lap.

Sully kicked at the main supporting branch and the roof sagged. The overhang collapsed, trapping Sully's leg under several feet of ice and frozen mud and created a snow-and-earth wall between the grenade and the ARCS team.

What followed was an explosion, louder than anything Taylor had ever heard. It blasted past her sending ice projectiles into her face. She looked up and saw the Denali storm silently raging above her. Where Sully had been, there was only a smoking pile of ice and snow.

Taylor opened her eyes and saw Dylan's face floating above her. His mouth was moving, but she couldn't hear anything he was saying. She yelled that Sully was buried behind her. She saw Dylan jump down and begin digging Sully out of the debris pile.

BESIDES TAYLOR'S DEAFNESS and Sully's terror at being buried, Dylan thought they both appeared to be in pretty good shape. The same could not be said for what was left of the soldier. Dylan dug through the remains of his pack to search for a weapon, but all he found of use was a coil of excellent climbing rope, a wicked-looking combat knife and two jambalaya MREs.

Through a now blinding blizzard, Dylan roped Sully and Taylor and himself together giving them 10 feet of slack between them. Remembering the route to the mining shelter paralleled the riverbed, Dylan led the remains of his team toward, what he guessed, was the mining shelter. Despite the shelter being only a mile away, it took them nearly three hours to get there.

Darkness closed in, the icy air shrieked malevolently and the party appeared exhausted from fear and shock. Dylan knew he had to get them rest and food. Oddly,

Sully had not complained about anything since the attack.

The shelter was not what Dylan expected. What he found was a shiny new prefabricated steel building that had undoubtedly been put up within the last couple of years. Snowdrifts several feet high leaned against the frozen walls and a heavy steel padlock hung on the door. "Sully, where's the key?"

"The key?" Sully looked dazed. His eyes roamed over the massive rusting hulk of a turn-of-the-century steam engine covered in rust and snow. "That's a big engine. It generated over 45 horse-power, too."

"Sully, we need to get some shelter. Where's the key to the padlock? You were here last summer, where did they keep the key?" Dylan had to yell over the wind.

Sully slowly turned toward a steep rock wall and gestured. "It's in there. I'll get it." He started toward a wall surrounded by huge snowdrifts. With Dylan and Taylor following, he walked between the two tallest drifts, grabbed a piece of plywood that Dylan had not noticed and pulled it aside revealing the opening to the old coal mine. "In here. There's a wooden office building constructed inside the mine. The keys are in it."

In the blackness of the mine opening, Sully reached down into a box and grabbed a handful of plastic tubes. He bent one and shook it. It quickly started glowing a pale green. "Grab some of these," he gestured to Dylan and Taylor as he pulled the rickety door closed behind them. They each put a dozen or so of the six-inch plastic tubes into their pockets.

Taylor activated a glow tube. "Sully, can we leave the door open? I get nervous being in a closed place."

"You're shitting me. Why did you come on this trip? You knew we were going to be in a mine." Sully shrugged and left the door open a crack.

Once inside the mine, the wind noise and sharp cold abated. They followed Sully's green glow down a gravel narrow gage railroad bed with a ceiling just high enough to allow the rail cars to move about.

They stopped at an old sagging wooden structure built against the cave walls with broken out windows and a floor littered with debris. Behind the door hung several new looking keys as well as some that looked decades old. He grabbed a shiny key and gestured for them to follow him back to the mine entrance.

Stopping before the open plywood door, and in a voice that was too loud, Sully said, "You guys wait here until I get the furnace going. It's probably warmer in here than the shelter."

A blinding flash from outside stabbed through the night; quickly followed by a concussive boom. Dylan saw the opening to the steel shelter smoking and the backs of two armed figures in snow-camo suits watching it burn. One tossed a second grenade into the shelter and then the pair ran behind some rusty mining equipment and turned away from the blast.

Dylan felt the two soldiers were looking right at him.

52

SHIRLEY AHN HAD WATCHED Cheek pick his nose and flick the booger away. Even across the room, she wasn't sure, but she wondered if it had landed in her hair. Every time she left a room that Cheek was in, she wanted to take a shower. If she didn't love being the governor's secretary, she would have quit when Leola first hired Cheek to lead the re-election campaign.

She'd need to get her dress cleaned, too. She was wearing a simple black lace-collared dress with a wide, shiny leather belt, black ballet flats and silver jewelry. She thought of this outfit as her sexy librarian look, but if she had known ahead of time she'd be working with

Cheek, she'd have avoided anything sexy or expensive to clean.

He looked up from his desk, "Shirley, can you look at this?"

"Shouldn't you be asking Mrs. Blodget, your own secretary?" said Shirley. *Oh crap. He wants me to come close to him. I hope he doesn't try to touch me.*

"She's out today," Cheek held out a paper, "Can you run this draft contract by legal? I think Jane Koss is going to join our side, and I want her locked in so she can't change her mind."

Standing as far away from him as possible, Shirley took the paper, careful not to touch it in the same place where Cheek' booger finger left a smear. "So you think she's going to change her mind and join our campaign?" *Make conversation so he won't try to pat my arm.*

"Frankly, I don't give a flying fuck if she does or doesn't. It'd be nice to have her talent and the propaganda value of a rat leaving a sinking ship," Cheek said.

"OK. Anything else?" *Please. Nothing else.*

"Here's a list of possible press responses from the governor when Bolton's 501 collapses. It's a very high priority that we get a quick indictment from the grand jury if we can move it that far. I emailed you the legal docs, and I've given a heads up to Judge Stewart."

"OK. Anything else?" Shirley asked.

"Yes, have you heard the governor's Mountain Club speech? I know she practices on you."

"Yes. I'd say she's ready," said Shirley.

"Did she use the speech I wrote? I'm especially interested in where she notes her opponent's opposition to any government regulation on oil exploration or carbon reduction activities."

"That's all in there," said Shirley. "I can send you what she has loaded into her teleprompter."

"Just tell me, does she mention that Yeung doesn't believe in climate change? Does she directly say that he views climate change as a political phenomena and not a scientific issue?"

"That's all in there. Anything else?" It would have been clear to anyone that Shirley was trying to get out of the room, but Cheek couldn't read people despite his amazing political mind.

"Yes, come over here and look at these press releases. They are to go out right after the Mountain Club speech. We would expect these allegations of fraud on the part of Yeung's 501 to hit the press right before that little Asian rat speaks before the Arctic Policy Board in two weeks." Cheek laughed when he thought of the chaos it would cause in the Yeung campaign to be hit with fraud right after he was socked with *the Scrubber* presentation.

Shirley stiffened at Cheek's "Asian rat" comment. *That bastard doesn't know I'm half Korean?* She didn't say anything to Cheek. She stood away from him and just looked at the press releases from across the room.

"Isn't it true that those allegations are just made up based on hearsay? We don't want to get involved in any prosecutorial misconduct. The governor won't hear of it."

"Relax. This stuff only looks like it comes from the Attorney General's office. There's enough wiggle room for him to deny everything. Since he's out of state right now, it will take some time before the truth comes out. Since we're not telling him about this, he can sound like he really means it.

"When Yeung finishes his little Arctic talk, all the questions will be about his legal troubles. Then the press will focus on his climate change views." Cheek laughed showing too much teeth. "That little slanty-eyed rice picker won't know what hit him. First our glorious leader will have announced an Alaska-based solution to greenhouse gases and then press will hit him with the fraud questions."

Shirley wanted to leave, but was curious. "Does the governor know what you are up to? She's said many times she wants a clean campaign."

"This is clean. Sort of. There really is no solid evidence. But we are trying to win an election. We don't need proof of our fraud allegations. We only need the press talking about it. Also, tell the governor she knows nothing about the fraud. If asked, she should say it's a surprise to her. That will get voters thinking the allegations didn't come from her campaign."

"OK. Is that all?" Shirley was paid by the state of Alaska. She hated doing campaign work while on the public payroll. She made a note to tell Leola to stop letting Cheek use her as a go-fer when his own staff was unavailable.

"No. I want a crew to tape Yeung's talk. That will make great video for a campaign ad if we need it.

"OK. Is that all?"

"That's all." Cheek started hunting for another booger just as Shirley left the room. "For now."

53

INSIDE THE MINE, Sully sprinted back to the office structure and ducked in. His boots crunched over broken glass as he pushed past a sagging door to an inner office. "There's a closet we can hide in."

"No! Sully, they are right behind us. We need to hide where they can't find us," Taylor pleaded.

It was obviously very hard for him to abandon his closet hiding place, but in the eerie green light from the glow sticks, Sully could see that the faces of his two friends were serious. Suddenly his face brightened. "I know where we can hide. Then I want Taylor to look at my leg. It's just killing me."

As they walked uneasily in the old mine, Taylor wondered if there was any place darker than a coalmine. The low ceiling seemed to press in on her and make it hard to breathe. Soon they heard water running and found a small stream flowing out of the wall. The water looked pristine. "I'm filling my canteen," said Dylan.

"Don't. Nearly all the water in this part of the mine is loaded with arsenic."

"Really? Sully, hand me your water bottle. I have an idea," Taylor said.

"My canteen is back there in Ethan's pack."

"Here's mine." Dylan passed her a half-full water bottle.

Taylor added Dylan's water into her canteen then filled Dylan's with the bad water. "Let's leave them a present," said Taylor holding up the canteen filled with bad water. "They may also be thirsty. If they drink this, it won't kill them instantly, but it could cause night blindness and diarrhea."

They followed Sully down the tracks to where a small tunnel branched off to the left. Taylor held up her glow stick looking down a side tunnel. "You guys stay here." She walked down the tunnel just to where it turned and dropped the glow stick and the canteen filled with bad water. Then she came shuffling back in the inky darkness.

"Great idea!" Dylan whispered. "Sully, now where?"

"Let's go get the coal we came for. It's only a couple of miles from here, and there's great hiding places back there."

From behind them, they could hear voices and see some light flashes.

"I bet that's Thing One and Thing Two. Let's get moving. Sully, you lead," Dylan said as he shoved his glow stick into his jacket.

"Taylor, you need to keep your body between your glow stick and those Things behind us. Or put it in your jacket so they can't see it."

"Walk on the outside of the tracks, there's fewer trip hazards, but watch the low ceiling," Sully said as he started down the tunnel.

After about 30 minutes of sloshing in the arsenic water past dozens of side tunnels, Sully held his glow stick up against the wall revealing a small, unremarkable side tunnel. "Here it is," he said. "Our first hiding place."

He turned left then walked though a brief maze of low ceilinged tunnels and came into a room carved out of the soft, coal walls. There were benches and lockers along the wall, as well as some empty wooden cabinets.

"Whoa. What was this room used for in the olden days?" Taylor wanted to know.

"I don't know. Storage I guess," said Sully. "Now for some rest and Jambalaya MRE. Oh, and Taylor, you need to look at my leg." With that, he started another glow stick, unpacked his MRE and gestured to Dylan for some water to activate the heater for his meal.

"We eat cold MREs. There's only about two liters of water for three people. We should be drinking 10 ounces of water every 20 minutes for proper hydration."

Taylor held her glow stick up to Sully's leg. "Holy shit! You have been bleeding! What's with your leg?"

"It started to hurt when that state trooper blew himself up. My leg was fairly near the blast. God it hurts like the devil right now. Can you fix it?"

Taylor opened her first aid kit. "It looks like a shrapnel wound. We need to disinfect it, but all I have is a bacitracin cream. We should flush it before putting on the cream."

Dylan took the dead soldier's combat knife out of his pack and used it to slice open a glow stick. A liquid ran out. "This is a chemical that reacts with hydrogen per- oxide."

"It's diphenyl oxalate I bet," said Sully through a mouthful of the MRE. He pointed to a glass vial visible inside the ruined glow stick. "It that what I think it is?"

"Yeah, it's hydrogen peroxide. This makes an excellent antiseptic wound wash."

"Do you think I can use my flashlight in here?" Taylor pointed to Sully's wound. "These black walls absorb light. Thing One and Thing Two would probably not be able to see anything from the main shaft."

Taylor used several glow sticks to get enough liquid to wash out the wound. Then she applied the bacitracin cream and bound it all up with tape. She gave Sully some analgesic pills to control the pain, switched off her flashlight, and then lay down on a bench. "I'm beat."

"Why are the state police trying to kill us?" Sully spoke through a mouthful of food. "I mean, we're just scientists. It can't just be for my data recorder. They could order one for a couple of thousand dollars. That trooper from ARCS asked for my data recorder before he blew himself up."

"I don't think these are real state police. In my experience, Alaska state troopers are well trained and professional," Dylan said. "They don't just kill someone unless it can't be avoided. These fake troopers are also really well equipped. You should have seen their campsite. There's money behind them. I'm guessing they are in bad shape now that their supplies are gone."

"Right now, I think there are two guys after us. That one guy blew himself up. He seemed almost eager to die for what he called, 'his cause'," Taylor said stretched out on a bench with her eyes closed. When she closed her eyes, the mine seemed less oppressive.

"So they want us dead and they want Sully's data recorder, and are willing to kill and die for it. It doesn't make sense," said Dylan.

Taylor spoke through closed eyes, "I want to know how Sully can keep a map of this mine in his head."

"I have a photographic mind. I was here most of last summer, and I remember the place. I'd really like to be able to forget stuff. Like I'd like to forget that Ethan's neck blew up. I'd like to forget the moment of his death. He was the one guy who didn't hate me. I'd really like to forget Zhang's face," Sully spoke through a mouthful of Jambalaya.

"We don't hate you. You are a hero. You got the guns away from that bad guy and you collapsed the river bank saving our lives."

"That river bank deflected the grenade blast back away from us. We got lucky on that one," Sully said. He stopped chewing, "You really like me? You really think I'm a hero?"

"Sully, you are a hero even if you are annoying at times," comforted Taylor.

Dylan lit the little room with what was left of his cell phone light. He looked over at Sully and noticed that he appeared like he'd been rolling in a bin full of coal dust. Taylor was also blackened by coal dust so that she looked like she had raccoon eyes. At that point, Dylan decided he must also look the same.

"OK, how do we get out of here once we get that coal? Those men are behind us," Dylan said.

Sully noisily chewed his crackers, "There's only one way in or out. We go out the way we came in."

54

JANE WALKED CONFIDENTLY into the Netsvetov Hotel's impressive lobby. She strode right to the penthouse-only elevator where a hotel employee stood in front of the ornate metal doors.

Dressed in the standard blue blazer and white shirt, the painfully slender young man smiled at her. "Good evening Ms. Koss. The penthouse elevator is closed tonight due to a private event."

"But my office," Jane gestured helplessly.

"Yes. The office suite is closed as well, and we have strict instructions that no one is allowed to use the elevator. I actually don't know why they are paying me to stand here. The elevator is locked at the top. The call button is disengaged." He shrugged.

"I guess I'll just have a drink down here in the bar." Jane echoed his shrug and approached a large bank of elevators near the registration desk. She pushed the button for the highest floor possible and scanned her card. She hoped that since her office was located in the office suite just below the penthouse, she could take the elevator as high as possible.

Once at the floor below the offices, she approached the stairs to walk up to the office suites, but she could see through the small window that a multiracial group of large men were in the stairwell smoking and playing dominoes. She opened the door, and they all stood up as if hyper-vigilant.

A large, very dark-complected black man wearing a black sidearm on his belt spoke to her with a Jamaican accent. "Hey Kid. This stairwell is closed. If you need help, I can call hotel security. Are you looking for your mom?" Other athletic-looking men silently evaluated her as to the threat level she posed.

"That's alright. I think I'm lost," she closed the door and pushed the button for the parking garage.

Once in the parking area, she entered the stairwell, and took the walk down to the vast kitchens and laundry area. Jane boldly followed a hotel employee into a *staff only* door and walked purposely to a break room. There she nodded at several men in coveralls drinking coffee and reading a well-wrinkled edition of the Wall Street Journal.

One of the men, a young guy with a blond beard watched her. She couldn't tell if he thought she was cute or wondered if she belonged there. Jane hoped she didn't reveal just how nervous she really was.

Once in the back of the dreary break room and out of sight of the men, Jane opened locker after locker until she found what she was looking for, a blue blazer that identified her as a customer service associate. It was much too big for her, but she thought it was better than going as a hotel guest. People wearing these blazers were invisible.

Next she entered the dark hall to the kitchen where she could see a set of four chrome trolleys lined up with room service orders waiting to be delivered. She picked the one with a tall coffee carafe and a large-domed catering tray. She wheeled it toward the service elevator hoping no one would stop her. Once inside, she marveled at the size of the elevator car. It was surely big enough to fit a king-size bed and then some.

Jane pushed the button for the penthouse office suite. On the way up she lifted the wide metal lid of the food dish. A strong fishy smell rose from a huge pizza covered in a grotesque amount of cheese and anchovies. She felt ill.

As the elevator rose, so did her anxiety. This was a bad idea. The Colonel expected his orders to be followed to the letter. He was *The Colonel*. But surely he would want her to interrupt his meeting to learn about Cheek's plans. Just as she thought she should abandon her efforts to talk to Bolton, the service elevator opened to the back entrance to the office suite.

A large Latino-looking man stood in front of the elevator door as it slid open. He looked like he was about to draw his weapon.

"Did you already get your snack?" Jane asked with a big customer service smile on her face. She needed to ask him questions before he could ask his.

"Snack? No one told me about no snack," he eyed the dome hungrily.

"Oh, the Colonel always has snacks for security help," Jane hoped he liked fishy pizza.

He lifted the dome and a puff of strong, fishy steam rose upwards, forcing Jane to lean away. The man smiled. "This looks amazing."

"Good. Now I need to go into that outer office to check for dirty dishes. I won't disturb the meeting in the office lounge."

The *Office Lounge* was, Bolton's name for the multiple couches, 50-inch digital TV and fireplace studded living room that adjoined Bolton's office. Jane turned away from the entrance to the office lounge and walked toward her modest office door. She took her Netsvetov Hotel key card and scanned the lock to her own office. A security light glowed green. Seeing that her card legitimately opened the door, he turned his attention towards the grotesque pizza.

As the service elevator door closed, she suddenly realized her key card would not call the service elevator back. She was stuck on the penthouse office suite surrounded by armed guards. Now she had to get the Colonel's attention without breaking up the meeting. *This was a bad idea.*

55

TANGO NUDGED THE WATER BOTTLE with the silencer at the end of her Weatherby Vanguard SUB-MOA long-range hunting rifle. At 50 inches, it was the longest gun of its type she had ever used. Now that they were in the mine, she wished for her Sig. Nothing beat it for short field combat.

They stood over the water bottle and glow stick. "Now we have some extra water." Boston bent down to pick it up.

"It's probably poisoned with something. I've tracked Dylan Baker before. He leaves surprises for trackers. Watch out for booby traps."

"You've tracked him before? How did he get away?"

"He didn't get away. The mission ended before I could kill him. Afterwards there was no point."

"Too bad Concord didn't get away. At least he died for the cause."

"Concord fucked up. He did not need to blow himself up. He was the one who had explored this mine. Now all we have are his hand-drawn maps." Tango held up a small tablet computer with a crude map on the screen. "I don't want anyone else falling on his sword for the cause unless I order it. Is that clear?"

"Yes, Ma'am. What should we do now? Follow this tunnel to the end to find our targets?

"No. This map indicates there is only one way out. We need to clear these side tunnels one-by-one and work our way to the end of the mine. From now on, one of us needs to stay in the main tunnel to prevent them from slipping past us."

"This is a big mine. How long will it take?"

"If we don't find them in ten hours, we're going to go to the entrance, start a mine fire and seal up the place. Since our transportation and supplies were destroyed, we need to leave ourselves enough resources to return to the Zetros."

"Dying in a mine fire seems a terrible way to go," Boston said. "There's no way they can escape."

"You're right, but I'd rather have a confirmed kill than a presumed kill. These long rifles are of little use down here. Cache your long gun, keep the ammo with you and use your sidearm. Our unarmed targets will be dead in ten hours or less."

SULLY SUPPRESSED A SCREAM when he attempted to stand. "God, my leg! What did you do to it?"

"You were probably in shock earlier. Now you are just aware of the pain," Taylor said.

"We can't wait here until he gets well. Let's grab the coal samples and determine a way out of here," Dylan said.

Cautiously they made their way back out to the main tunnel and began to walk the tracks using only one glow stick for the group. With Dylan's cell phone dead and Taylor's nearly so, the glow stick was precious.

Dylan carried a six-foot long crowbar he'd found. It wasn't much of a weapon, but he felt better when holding it.

Sully's limp became more pronounced, so Dylan found himself having to support Sully's weight on his shoulder. Not only did this slow them down, it made their progress much more noisy.

So Sully stopped. "I can't go on any further. Just stash me in a side tunnel, cover me with coal and pick me up on your way back."

"No way. You're the only one who knows where we are going. Plus, they could find you, and then you'd be dead," Dylan said as they rounded a sharp bend in the rail tracks.

Taylor dug around in the first aid kit for some more powerful pain medication and handed a coal-dust covered pill to Sully.

"How about hiding me in this old coal car," Sully swallowed then gestured to a bathtub-sized iron coal cart lying on its side near the tracks.

"Hey, if we can get this heavy thing up and on the tracks, we can push Sully," Taylor said.

"That's a great idea. It could save me about a mile and a half of agony. This coal car must have tipped off the tracks as it rounded the turn. There's another turn about a mile from here. We'll need to make sure it's not going fast when we reach it." Sully winced.

Using the crowbar and some broken track, the three managed to get the cart back on the tracks, but the wheels seemed frozen with rust.

"Pee on them," said Taylor. "I'd do it, but I lack the right equipment. Urine has uric acid and other nitrogenous wastes. It should help dissolve rust and provide lubrication."

The men provided the proper acidic lubrication and, with an ear-piercing screech, the cart began to move forward. Sully settled himself in the cart with his arms and legs spilling out while Dylan and Taylor walked beside the cart each with a hand on a rusty iron handrail.

FROM FAR AWAY the shrill shriek of the cart's wheels pierced the darkness. Tango grabbed Boston's arm. "They're in the main tunnel on the tracks. No need to search side tunnels as long as that squeaking continues. The two checked their pistols and started a light jog down the tunnel with LED flashlights blazing. Tango smiled. She knew she'd get the confirmed kills and be back at the Zetros by dawn.

56

ONCE IN HER OFFICE, Jane started worrying about how much time it would take the outside guard to raise the alarm. She had told him she was looking for dirty dishes. Once his huge pizza was gone, he would notice his own dirty dishes and figure out she was not back yet. Jane had to move fast. But what to do next? Briefly she wondered if she should just hide in her closet until morning.

She knew it would be a mistake to just barge into the meeting and ask for a breakout meeting with the Colonel. He had taken extraordinary measures to make sure the meeting was private. She would need to lure him out of the meeting somehow.

Growing up with four brothers, she had often played hide and seek. She knew how to open doors quietly, hide in the open and use subterfuge to hear their plans for finding her. First she noiselessly approached the door that adjoined Colonel Bolton's office with her office and listened. It was silent and an index card pushed under the door did not reflect any light. The lights in his office were off.

He was always barging his way into her office from his side without knocking, so she had taken to locking her side of the door. She hoped his side was unlocked. As she turned the lock on her side, it made a loud metallic *thunk* that startled her.

She became still and held her breath for what seemed like minutes. The hammering in her heart slowly began to subside. She opened the door on her side, and she gradually pushed against the door on Bolton's side. It barely moved. Something was in front of it.

She pushed harder causing the door to open a few inches. She looked down to see a file box full of direct mail advertisements. *Why would he leave that there?*

Using her phone as a flashlight, she noticed the Colonel had also left a metal vacuum bottle next to the box. His special tea, he took it everywhere. If that metal bottle fell on the hardwood floor, it would clang like a church bell.

Soon she had made her way through the door and out into the Colonel's expansive office. She noticed his laptop open on his desk. He was so paranoid about security; she had never seen it open before. He closed it whenever she entered the room as if he was concerned she would learn secrets by glancing at the screen.

Curious, she approached the glowing laptop and noticed the screen was set to the controls for the complex set of security cameras he had installed all over the penthouse and office suites. It looked to Jane like he had set the recording to OFF to prevent any record of his meeting.

She could see a light under the door leading to the office lounge. As she approached the door, she could hear the murmur of male voices. She would have had to open the door to clearly hear what they were saying and possibly get the Colonel's attention.

As if it were the minute hand on a clock, she slowly moved the doorknob praying that there would be no squeak or click or that any of the occupants of the room would notice. Once the door was open a crack, she could see six men sitting in the glove-soft leather office lounge chairs.

She immediately recognized Bolton's CFO. Bolton's ever-jittery CFO/chief accountant was looking like a nervous cat. He sat among the other men with his freckles and bow tie, twitching all the while.

From her vantage point and the positioning of his cell phone, it looked to her that Bolton's CFO was recording the meeting on his phone. She needed to inform Bolton of that. What if the CFO might be gathering evidence if he decided to someday testify against Bolton. She could restart the Colonel's security system and show him what she was seeing.

She returned to the laptop and clicked the button that restarted all the cameras. Jane tiptoed across the room back to door and looked out. One of the men, a literal caricature of a Middle East oil prince, wore a pure white robe with gold embellishments running through it.

On his head he wore a white silk scarf held down with a thick, doubled black cord that matched his trim goatee and a wickedly curved dagger on his belt. He spoke with a British accent. "But Colonel Bolton, how do you know you can trust Greg Yeung to open the wildlife preserves and coastal areas to drilling?"

"That's no problem. The little asshole is in my pocket. I got recordings of him saying he's an atheist.

He'll do whatever I ask." Colonel Bolton swirled his whisky confidently as he spoke.

A very small, meticulously dressed black man spoke in a soft voice, "We are concerned about the market for oil. With all the fracking in North Dakota and the flood of oil in the market, how do we know the market for oil will be reliable? Don't we already have enough oil?"

Bolton stood up, "That's the sweetest part of this deal, and we are close to getting a federal law passed for stricter limits on coal-fired power plants and fracking because they are so dirty.

Fracking floods the atmosphere with methane as a by-product of natural gas extraction. As soon as our legislation goes through, most of them will switch to our non-fracted natural gas or diesel-power. Oil will once again be on top. Yeung will make it nearly impossible to mine coal in Alaska so that's a start. Other states will follow. Banning coal is the new gay marriage. It's going to happen."

Another man, dressed in a cheap suit with no tie, spoke with a South American accent, "All this seems too good to be true. What are you putting in? You promised to match our contributions."

"That's right," the Saudi Prince said. "You don't have the capital to match our bids on those drilling rights."

Bolton rubbed his hands together, "I know where to get the money. All I need to do is plant a story that Exxon is buying out my company and my stock price goes way up. Once the price peaks, I sell. Afterwards I plant a story about disappointing earnings and the price falls. I buy. The public are fools. I'm still taking a bigger risk that any of you."

The nervous CFO looked aghast at Bolton. It was clear he did not know about this latest stock manipulation. He burst out, "But Colonel! Our cash flow problems! You leveraged all your companies to buy those drilling rights! We have bills. We need cash."

"Relax. It's all arranged. As of this meeting, the Super PAC has more than enough to cover our needs. We borrow short-term from the Super PAC and pay it back before anyone is the wiser." Bolton dismissed his CFO's concerns. If possible, he was even more edgy.

For the first time, a tall, pale man in an expensive suit spoke with a harsh Russian accent, "According to the map you gave us, much of the exploration is to be done on Native American lands. How do you know they will allow oil exploration?"

Once again, a question rekindled Bolton's excitement, "Those fucking Indians are so stupid and trusting. All we need to do is tell them they'll make more money than a casino, and the elders will cave in. It always works."

"Are you sure? They've been turning down deal after deal saying their lands are being harmed by climate change. What makes you think they'll change their song? "

"Change their tune," corrected the slender accountant-looking man. "Viktor is right. What makes you think they will see it our way?"

"Look, they are all pushed out of shape because they can't do their winter hunts due to the mud. It used to always be frozen and it was a piece of cake for them to get their winter meat. All we need to do is show them how much meat they can buy with their extra money." Bolton poured more drinks for everyone.

Jane stood in stunned silence. *Native lands, offshore drilling, wildlife preserves and more stock fraud! That arrogant asshole was risking 20 years in prison and she might get pulled into it!*

Now she felt trapped. Unable to leave, but unable to move forward, when her boss was so obviously corrupt. And Greg Yeung, an atheist? Why would he use the word, *blessed* so often? If this got out, his political ambitions would be toast.

As she backed up toward her office, she knocked over the metal vacuum bottle. It fell with a ringing crash and rolled noisily away.

57

THE PAIN PILL TAYLOR had given Sully seemed to be working. He stopped complaining and started cheerfully babbling while his coal-dust-covered white-framed glasses lay crooked on his face. His coal-cart ride sped up to a fast walk for Dylan and a light jog for Taylor.

The going was tricky because the ceiling was low. Dylan, not an especially tall man, had to duck to avoid contact.

"What color do you think this coal is?" Sully gestured grandly at the tunnel.

"Black. It's coal," said Dylan.

"Nope, this is brown coal. This is the shit in the last test burn. It's full of arsenic, sulfur and heavy metals. It's only about 55% carbon and nasty. It was used as a fuel for steam powered boats during the gold rush years, but it caused so much damage to the boilers, that boat owners hated it."

"So why did they use it?" asked Taylor.

"Because it was available. Thousands of miners wanted to get around and the only summer transportation was boats. Boats needed coal. This mine is only about 100 miles from the Yukon River." At first Sully was bellowing to talk over the squeaking wheels of the cart, but the squeak had settled into a quiet grinding sound.

"The coal we are after is quite different." Sully slurred his words. "It's from a seam that used to be deep in the earth, but all the geological activity near Denali has moved it up."

Dylan motioned for Sully to whisper and pointed back where they came from. "So this coal we're getting isn't brown?"

"It's not really coal. It's like an ultra-high-grade anthracite. The purest I've ever seen. It has the highest calorific content of any coal ever tested, and it's in a seam the size of Dallas and miles deep. It's nearly 99% carbon, and if we can extract the carbon dioxide from the combustion gases, it can supply the energy needs of the five western states for the next 25 years without any greenhouse gases. Now all this is classified so don't tell anyone," Sully started laughing.

"I think he'll get only half a pill next time," said Taylor over Sully's chatter.

"Hey, the coal we are after isn't even dusty. You can rub a white glove on it and get no residue. It shines like glass," Sully giggled.

"When do we get there? I'm tired of all this jogging in the dark. At least the tracks are heading downhill."

"We should stop this cart in 20 minutes then walk the last part. It's too steep to use a cart without brakes." Sully closed his eyes.

Suddenly Dylan could see the white beams from LED lights on the ceiling in front of him. He looked back to see two painfully bright flashlight beams jogging down the cart tracks.

A sharp spray of sparks erupted from the back of Sully's coal cart and a loud boom shocked their ears.

"They're shooting at us!" Dylan yelled. Sully became fully alert when another three booms in quick succession echoed through the mine. A fine mist of coal dust fell from the ceiling and walls, filling the already dusty air with a thick mist.

"Dylan, if you ram your crowbar into the ceiling, you can bring down some coal. They'll have to climb over it or dig it out to get to us," Sully pointed upwards.

As Taylor and Sully continued rolling, Dylan stopped and found a crack in the ceiling. He rammed the bar upwards. It penetrated much farther than he expected. He couldn't budge it. Pulling it out a little he tried again. This time he saw the muzzle flashes as three more shots were fired. He pulled once more and this time he had to dive away from a fall of rocks that nearly filled the tunnel with debris and dust. Coughing, he ran to catch up with Sully and Taylor.

Looking back, Dylan could see the lights of his pursuers as they moved the soft coal lumps out of their way.

"We only have one chance," yelled Sully. "Jump into the cart."

Taylor threw herself across Sully's lap, her legs hanging out of the cart. Dylan leaped onto the back of the cart forcing Sully to lean over Taylor. Without anyone holding on the sides as a brake, the cart began to pick up speed and rock back and forth dangerously.

The lights of their pursuers disappeared when the tunnel made a gradual turn, but the speed of the cart increased as the tunnel became steeper and the darkness became complete. As the wind whipped Taylor's hair into his mouth, the clattering of the cart rattled Dylan's teeth. Dylan thought of all the bad things that could happen on the next black turn.

58

AFTER 20 MINUTES hurling through the inky black mine with no idea of how fast they were going, what lay ahead, or what would happen when they reached it, Dylan began to know pure terror.

He felt small comfort at being with Taylor and Sully. The bearings of the coal cart were burning up and screaming their agony through the mine with an ear-splitting noise. Taylor was silent and Sully was whooping as if riding a theme park water slide.

Dylan knew that behind him were some killing machines that probably knew there was no way out of the mine. His main contact with reality was the long steel crowbar in his hand. He looked down at where the cart's wheels spun madly below him and got an idea.

Pushing the high carbon steel of the crowbar between the hot, glowing steel wheels of the coal cart and the track, Dylan sent an amazing spray of sparks into the air just behind the cart. In no time, the tip of the crowbar began to also glow red, but it was probably hardened steel that could get quite hot and still retain its shape.

Soon the steel transferred the heat up the bar so that Dylan had to pull off his sock hat to use it as a mitt to hold the steel bar. Still sparks spurted out behind in a rooster tail of harsh light and the coal cart hurled through the darkness.

Sully laughed and handed Dylan a length of rusting chain he had found in the cart. Dylan fed this into the other wheel, hoping a bump of the coal cart wouldn't toss him off.

Another spray of sparks spurted off this wheel, and Dylan could feel the cart start to slow, but it was too late. A 30-degree turn threw the cart and its occupants into the darkness.

The world tumbled and all became black and still. Dylan coughed. From about 20 feet away he could see the dying glow of the cart's hot metal wheels high on the ceiling. Somewhere in his pack he found his flashlight and switched it on revealing a thick coal-dust cloud hanging in the air. Then he figured out he was upside down on a tall pile of coal lumps.

"I'm not dead! I love this soft brown coal. It's more like heavy styrofoam than rocks," Sully drawled drunkenly. Dylan could see he was partly buried in the coal. A cut over his eye was bleeding, but not too bad. Taylor was likewise buried, and disoriented.

"We need to get out of here. Those guys are coming." Dylan tried to convey a sense of urgency.

"There's no way out," Sully said. "I vote we pull a pillar and bury the bastards. Then try to dig our own way out."

"Pull a pillar?" Taylor asked.

"This old mine was worked with a room-and-pillar approach. The miners dug long rooms through a coal seam leaving some parts unmined. These parts would be the pillars that held up the ceiling. When all the rooms were cleared out, the minors would remove the pillars causing the ceiling to fall in, then that rubble would be mined."

Sully pointed to a floor to ceiling pillar of coal about 10 feet on a side. "The pillars are usually about 100 feet wide. This one is ready to pull."

"OK. So our choice is being buried in a mine under tons of nasty brown coal, or being shot by some bad guys. Do I have that right?" Taylor wanted to know.

"That's it," Sully slurred. "Except the anthracite coal seam is only about 400 yards away. Last summer we had some tools that we used to dig out and drill core samples. We can collapse this ceiling, then use the tools to dig ourselves out. We have a chance."

Dylan could see the logic. "How do we pull this pillar?"

In just a few minutes, the three had fetched the sampling equipment from the anthracite coal seam and rigged the battery-powered drill as a winch.

Sully was more in the way than helpful, but soon the improvised winch looked ready. Dylan had wrapped the rope around the pillar and back to the winch. "How do you know there's enough battery power left in that drill to winch the pillar down?"

"I don't know. I can only hope. You might need to shave off a few more feet from the pillar to allow it to be pulled." Sully set down the water bottle. "Looks like we're out of water now."

"God damn it! You drank up all our water?" Taylor was livid.

Just then a pair of LED lights stabbed through the darkness revealing a stunning amount of dust in the air.

"Shit! They're here! Pull the pillar!" Sully yelled. He turned on their makeshift winch, and Dylan drove his crowbar deep into the pillar. He pulled out basket-ball-sized chunks of soft coal as the rope tightened.

Muzzle flashes from the pistols, quickly followed by painfully loud booms, gave Dylan more urgency in his digging. Sully turned up the winch speed to high.

The LED lights bobbed as the assailants raced toward their targets.

The winch made a grinding sound like it was in distress. The winding slowed down. Dylan tried to put the pillar between himself and the shooters as he dug.

Suddenly the rope snapped as it cut through the soft coal pillar. Dylan could see several feet of space between the top and bottom of the pillar, but the ceiling did not collapse. Two more loud, concussive shots rang out, and Dylan could feel the side of his jacket jerk. He wondered if he'd been hit.

From somewhere in the earth, a deep subsonic groan started. The noise was so strange and powerful, that all froze, even the shooters.

"Run!" yelled Sully as he limped toward the anthracite part of the mine. Taylor and Dylan were not far behind as the groan changed to a loud snapping and the sounds of thousands of tons of soft brown coal slamming onto the floor.

The dust cloud had followed them into the new part of the mine, but after lots of coughing, Sully could finally speak. "Well they can't get us now. It would take years to hand-dig from their side of tunnel to us."

"Years?" Taylor said. "So how long will it take us to dig going the other way?"

"The same," Sully said.

59

CONRAD LIMPED INTO Dave's office holding a bag of Sully's mints. "Hey Dave, during a lull in the storm, we picked up an ERB signal near the combustion station."

"What the hell? The away party must have passed that area hours ago, what do you think it is?" Dave put down his pen and stood up, towering over Conrad.

"Don't know. It's unlikely it could have gone off accidentally. The beacon was checked out to Sully. So he is either in trouble or it's a technical difficulty," said Conrad offering the bag of candy to Dave.

"It's really nasty out there right now," said Dave. He looked at the Halloween decorations on the bag and shook his head. "Those look like really old candies."

"Actually they aren't bad. You need to get used to the idea, and they taste fine," said Conrad. "So what do you think we should do? Send out another away team?"

"I can't ask anyone to go out in that weather," said Dave. "But an emergency beacon means an emergency. I'll see if I can find a volunteer and check it out."

"You can pull a sled with emergency medical supplies. That rescue toboggan out in the mudroom only weighs 25 pounds and it's big enough to drag a person back here," said Conrad through his mint.

"OK, I'm going to see if one of the techs will be willing to brave the storm with me. But we need to get back before the scheduled teleconference with the governor."

"I'll pack the toboggan," said Conrad. "You round up your tech and suit up."

HOURS LATER, Dave and Lindsey looked down at Zhang's body. "He looks frozen," said Lindsey her voice husky with emotion as she gazed at his contorted face.

"That he is," said Dave. "He must have come out here on his own and died from exposure. Dylan's away team probably found him. If he was alive then, they'd have returned to the base."

"That's why the emergency beacon was set, so we could find him later," said Lindsey. "Let's load him up and get him back to ARCS. We need to notify his family and make arrangements."

"We're going to leave him here. We'll put him into the toboggan and cover him with snow."

Lindsey looked shocked, "Leave him here? But, Professor White."

"We can't do anything to help him. When this storm abates, we can send the snow cat out for his body. Meanwhile, I have the teleconference scheduled with the governor, and we don't want the staff upset."

Lindsey hid her uneasiness under the hood of her parka. "OK. If you say so."

"It's for the best. I know you don't want to leave his body out here, but if we wrap it up in the toboggan and turn it upside down, animals will not be able to disturb it. Right now it's below freezing out here, so there's no danger of immediate decomposition."

Dave reached down and deactivated the ERB from Zhang's pocket. "Let's get started."

60

FOR SOME REASON, the knowledge that the shooters were either dead or on the other side of tons of dirty coal, did not calm Dylan. He felt anxious and eager to find a way out of the mine. It should have been quiet in the depths of the earth, but constant echoing drips and deep subsonic groaning from the area where they had pulled the pillar, set him on edge. He longed for his woods near Seward. He could tell that Taylor was showing signs of anxiety, too. He hoped she did not succumb to her fear of closed places.

Soon the dust began to settle. A hundred yards from the collapsed ceiling, their waning flashlight beam reflected off the nearly pure carbon walls of the anthracite seam and revealed a cavern with an apex nearly 60 feet above their heads.

The sharp but dimming LED light made the mine appear like their prison was a cut glass inverted bowl with thousands of facets reflecting light into uncountable points.

"This place reminds me of a crystal ice cave inside a glacier," said Taylor. "It's beautiful, but I want to get out of here. I can't breathe."

"Maybe there's enough battery power left in the winch to drill a hole through the fall. Then we can tunnel out," Sully suggested as he rubbed his leg. "Any more pain meds left?"

Dylan started their last glow stick and walked over to the collapsed shaft to survey the area. The winch was in bad shape, but the rope looked pretty good. Soon Dylan found Taylor and Sully standing next to him. "The flashlight finally died. We brought more flashlights, but they were in Ethan's pack."

"That glow stick is our last light except for my data recorder and our cell phones. It should have about an hour left in the batteries, depending on how long Zhang used it," Sully said.

It seemed like a long time ago that they had found Zhang's body. Now they were stuck in a mine with nearly no water and about to lose their last bit of light. Dylan didn't want to say anything about the ruined winch.

"Why is this mine making that groaning sound? It's creeping me out," Taylor said.

"There are many parts that still need to fall. Some-times, after a pillar is pulled, the ceiling will collapse over a period of months or years. Try not to stand under it. I wonder how much air we have in here?" said Sully. "I think we're pretty much sealed up in this mine."

"Aren't you cheerful," said Taylor. She looked at Dylan. "Do you have any idea how to get out or is this it?"

"If this is it, I'm going to eat that last Jambalaya MRE," said Sully empting his pack in the green glow.

"Really, that's a good idea. Let's take a rest break in the anthracite part of the mine. It's not so dusty there, and the ceiling looks stable. We can eat something. Then we can think of a way out," Taylor said.

Dylan agreed. Back in a clean part of the mine, the remaining rations were quickly divided up and consumed. In moments, Sully was asleep on his pack.

Taylor moved over next to Dylan. "Enjoying our outdoor adventure?" she whispered softly so as not to disturb Sully.

"Actually, this isn't my first trip around the tree trunk, and I've learned that my adventures don't always end badly. The last time I was chased and dodged death in the Alaskan forests I ended up with a new wife!"

Dylan gazed at the glow stick for a moment. "Accepting Governor Bates-Hardy's assignment provided me with the chance to build a relationship with my daughter." Dylan spent a few moments reminding Taylor about his first meetings with Anja and his efforts to create a father–daughter relationship.

"With Suzie's writing assignments constantly pulling her out of Alaska, I've been by myself. I find I miss having someone with me to share my peaceful life."

"Actually, in many ways my dreams are similar to yours. I would love to share a life in the woods with someone—if you add in a little science. Good night, Dylan." Taylor turned over and quickly fell asleep.

Dylan fell asleep before the last glow stick faded to black.

Hours later Dylan awoke. For some odd reason, he thought he could see the others sleeping. It should have been utterly black in the eternal night of the coalmine. Dylan wasn't sure, but he thought there was a faint shimmer coming from near the place where they had pulled the pillar. Without any flashlight or glow stick, he walked to the groaning soft coal seam and looked up. High up overhead, he could see a dim light.

"Wake up everyone," Dylan called. "I found something you're going to like."

A voice in the dark returned the call, "More food? More pain meds? More water?"

"Better. I found a way out."

61

TAYLOR, SULLY AND DYLAN stood under the groaning ceiling and stared up at tiny patch of weak light high above them.

The light emanated from a laundry basket-sized hole near the apex of the three smooth walls, which came to a point near the top of a domed ceiling. He observed that three sides of the gymnasium-sized room appeared to be smooth granite and the fourth was a pile of rough coal.

"Well, that's that. We can't get up there unless we had an elevator or a 60-foot ladder," said Sully lugubriously.

"Sully's right," agreed Taylor. "That's out of reach of us humans. Only a gecko or a fly could get up there."

Dylan studied the huge pile of rubble and sheer granite walls. Unconsciously he plotted a climbing route.

"I can do it," said Dylan. "I climb walls for a living. We have a rope. If I can get up there, rig some kind of pulley system from all this drilling stuff we have lying around, I'll be able to get us out of here."

A thin drizzle of water made a splashing sound on the floor under the light. Sully tasted it. "It's sweet, not bitter. This is snow melt."

"How could it be snow melt?" asked Taylor. We've gone deep underground."

"The surface above us is not flat," answered Dylan. We hiked up to the mine opening, and the shaft we were in followed the slope of the mountain downhill. We are close to the surface."

Sully turned, "I'm going to get my canteen."

"Fill all of them, Sully," said Taylor. I'll bring the rest of the gear in here." Just as she said that, a loud cracking sound snapped through the area followed by a fall of rocks somewhere.

"We should hurry. I think the rest of this mine is trying to collapse." Dylan started pawing through the drilling equipment that Sully's team had left there last summer. He made a pile under the opening.

"Hey, it's getting lighter in here. I wonder if it's dawn? I'm going to get my data recorder and check the time. My phone got smashed in that coal car crash." Sully retreated back to the where the packs were left leaving the open water bottles under the fresh water drip.

An hour later the party was assembled under the opening, which appeared to be about 60 feet above their heads. Dylan had taken off his coat and most of his outer gear. He stood in the pajamas that Anja had bought him. The rope was coiled at his waist.

Since leaving his home in Seward, he had felt off balance and out of his element. Sometimes he felt foolish because he didn't understand all the hidden political and scientific currents that swirled around him. Seeing a chance to climb flooded him with a calmness as well as filled him with energy and focus.

Instinctively, Dylan began oval breathing. Climb was what he did.

"Hey Dylan. Do you always climb in your pajamas?" Sully asked in a light teasing voice. Sully could feel a huge sense of relief that soon they would be out of the mine. "When we get out of here, I say we head south to a gold mining camp just a few miles from here. If Thing One and Thing Two got out, they might be looking for us on the route to ARCS."

"That's good thinking, Sully. I don't think those guys came from the south. They were tracking us right after we left ARCS," said Dylan.

After hydrating, Dylan planned his ascent. Instead of starting up the wall nearest the light, Dylan approached the groaning walls on the opposite side of the cavern. He scrambled up a pile of dirty coal lumps until he reached some crumbly, soft sandstone.

The soft sedimentary rock would be terrible for a technical climb, but Taylor and Sully watched Dylan edge along the sandstone to where it came in contact with a more solid, conglomerate sedimentary rock. He traveled this layer far from the opening he was actually aiming for, but he found what he wanted, some high quality granite folded against the soft sedimentary rocks.

Dylan understood and loved the rough, sparkly granite. Just touching it seemed to pour energy into him.

After about an hour of dangling from his fingertips, swinging his body like a pendulum and short rests, Dylan found himself at the opening. "OK, I'm sending down the rope. Tie the second bundle to the end. I can definitely make a pulley system up here."

Just a few hours later and 60 feet above the mine floor, Dylan had twisted pipes into the earth by hand and rigged up pulleys to hoist Taylor up and out of the mine.

"Whoa! It's really cold out here! And foggy!" Taylor shouted from outside.

Sully looked nervous. "Do you think that skinny rope will hold my weight? I'm over 200."

Dylan figured he was 250 but assured him it would. Dylan secretly worried about some unseen damage to the rope when the coal fell on it. He also worried about more unstable coal settling into the mine causing problems. He could hear small coal deposits falling and some very loud and eerie groaning.

After Sully and Taylor were out of the mine and complaining about being cold, Dylan, standing on the mine floor, sent up their packs and donned his own outdoor gear. The last thing out of the mine would be his pack hanging below him and now swinging from the end of the rope. Shivering, he started pulling himself up. Above, Sully and Taylor also pulled and got him within ten feet of the opening. Suddenly a deep rumble shook the ground under their feet.

Dylan felt weightless for a moment as he dropped down another 10 feet before Sully and Taylor arrested his descent. This time a sonic boom blasted Dylan from the mine as an interior collapse forced air out of the opening and a rush of black coal dust exploded up, buffeting and choking him.

Holding his breath, Dylan scurried up the rope hand-over-hand as the others pulled. Dylan rolled out onto a dirty snow bank coughing just as another, much larger blast of coal dust burst out of the mine in a huge, black billowing cloud.

As soon as he could catch his breath Dylan started laughing. Sully and Taylor looked as if they were made up in theatrical blackface makeup. Then he realized he must also look the same.

Sully looked at Dylan and Taylor covered in black coal dust and started to giggle. "We're all black," he said.

"Dylan, is your pack still down there?" asked Taylor as a sharp wind gust made her catch her step.

"Yep. It's probably buried under tons of nasty coal." Dylan looked into the blackness of their escape route.

"Oh no! I put all the water bottles in your pack!" wailed Sully.

62

SURROUNDED BY HIS HUMMING lab equipment, Professor Dave White sat in front of his laptop. He felt it made him look more professional to emphasize his science environment. And he needed all of the professional trappings he could muster.

After booting up his laptop, Dave spent a couple of minutes reviewing the Critical Issues Summary: High CO_2 levels, dead bacteria and data indicating the presence of toxic brown coal. Also, an unexpected power usage might be an issue, but maybe not. And an away team enroute to procure black coal free of toxic heavy metals.

He looked over at Conrad. "So is this link secure? I don't want anyone hacking into this call."

"We have an IETF encrypted media stream for VoIP and it requires a client-side certificate. Yeah, with end-to-end encryption, it's about as secure as an internet call can be," said Conrad just off camera.

Dave didn't understand all the industry standards for Voice over Internet Protocols, but he trusted Conrad.

"Who else besides my computer and the governor's computer can decrypt the call?" asked Dave.

"No one. Really, it's about as secure as any video-telephone system can get."

A chime sounded from Dave's laptop and a virtual *connect* button lit up. "Here goes," said Dave and clicked on the button.

IN THE ANCHORAGE Governor's Mansion, Governor Leola Bates-Hardy sat at the ceremonial desk facing her laptop. She would rather just talk at her regular desk, but Harold Cheek insisted she talk here.

If he weren't the most brilliant political mind she'd ever met, she would never tolerate him near her. She was turned off by his perpetually rumpled appearance and his whisky-and-coffee smell. It irritated her that he had swept her Dr. Pepper can out of reach so it would not be visible during the call.

She clicked on the *connect* button and waited for Dave's face to appear on her screen.

"Hey Professor," she said. "Can you hear me?"

"Yes, Governor. Loud and clear. How are you?"

"Fine. I understand you are having some storms. Is everything ok?"

"We're managing. Thank you, Governor."

"Tell me. Do you have good news?"

"Yes. We have our little surprises, but I see no reason for delaying your announcement to the world," said White.

"I can't tell you how important this is to me. The idea that we can burn Alaska coal without adding greenhouse gases to the environment just makes me giddy."

"Of course, Governor. Me too."

"I'm interested in learning how your team plans to get the Scrubber into every industrial exhaust pipe. Implementing that plan will go a long way to actually saving the world."

"It's a bit early to confirm that, but we can say it works for coal-fired power plants. And if we burn the pure coal we have here in Alaska, we won't have the other pollutants associated mostly with coal-fired power plants."

"I want you to convey my sincere congratulations to your ARCS team. Everyone there deserves Alaska's gratitude."

"Governor, I'll pass that on. Thank you."

"Thank you, Professor White. I just wanted to hear that we could proceed with our announcement. Goodbye, Professor."

"Certainly, Governor. Go ahead with the announcement. I'll send our most recent progress bulleted out in layman's terms."

IN THE ZETROS, Tango looked over at the soldier she called Concord, "Did you get that recorded?"

"Yes, Ma'am. We don't have a video monitor in that part of the lab, but we captured the audio."

"Good, encrypt it and send it to The Boss. Have you heard from Phoenix?"

"Yes, Ma'am. They buried Ethan's body in the mine near where we presume the others were buried in that ceiling collapse."

"Perfect. We want the bodies together in case they are exhumed," Tango thought back to her escape from the collapsing mine and shuddered involuntarily.

"Should I order the burial detail back to base?" asked Concord.

"Yes, but first have them set a charge to collapse more of the ceiling. We want it to take years before a search party can recover any of the bodies we left back in the mine."

63

"OK SULLY. Which way is this mining camp? We're out of food and water, so we need to get there soon," Dylan said. He worried about the state of fatigue he found himself in. He knew that if he felt exhausted, the others did, too.

"If we get thirsty, we can just eat snow. No problem with thirst," said Sully scooping up some snow to suck on.

"When you eat snow, your body expends energy warming up after eating the cold snow. Unless you have food, eating snow is an unsustainable method to stay hydrated," said Taylor.

"Damn! I get it. Now, to find the mining camp. When we went there last summer, there were about four miners and they had plenty of food and water. They had some good cookies, too. The kind that look like Oreos, but are lemon flavored. Just keep going south. They are only a couple of miles away," Sully said.

"Which way is south?" asked Taylor. She swung her arm to indicate they were surrounded on all sides by snow and fog.

"Your phone has a GPS. Just check it to see which way is south," said Sully.

"All of our phones are dead," said Taylor.

"I think I can get us on a southern route," said Dylan. The rivers around here flow west, so we stay 90 degrees to the left of each little frozen river we encounter. Does that sound right, Sully?" Dylan asked.

"When we were here last summer, we just used our GPS. I do remember we kept the mountain on our left as we traveled. We had 4-wheelers though."

Dylan thought about the GPS devices in his and Ethan's packs. He decided not to remind the small party about Ethan. In the enveloping fog, he knew it would take a long time to hike back and attempt to locate Ethan's ransacked pack.

After an hour of hiking Sully's limping became more pronounced, and his pace became slower. Any kind of elevation change required both Dylan and Taylor on either side of him to help him up or down. Dylan could see that Taylor's strength was fading. He hoped they would find the camp soon.

Dylan urged the others to keep small amounts of snow in their mouths to stave off dehydration, but at some point their bodies would run out of energy to melt snow and their core body temperatures would drop.

It seemed to Dylan that the fog was becoming denser. Their world shrank to just a few yards. Taylor seemed to be shivering. Dylan wondered if the party would be better off digging a snow cave and sheltering for the night, or to press on in hopes of getting to the camp soon.

"I need to rest," Sully announced with a voice that seemed weaker than anything Dylan had heard since meeting him.

"Great idea. We'll take a rest at the top of this rise," Dylan agreed. He wanted Sully to be facing a downhill walk when they resumed their trek. Dylan wanted himself to also face a descending slope. He was finding himself thinking irrational thoughts attributable to exhaustion.

Taylor walked beside Dylan. "I'm beat, too. I think I'm worn out because I'm sad about Ethan. I'm worn out because that adventure in the coalmine scared the hell out of me. And I'm hungry and tired. We need to stop soon." Taylor's chin bobbed with shivering.

Dylan agreed, but they continued to slog up a rise as the fog darkened and thickened. Dylan couldn't be certain which direction they were traveling, but he could tell the afternoon was changing to evening. Soon it would be dark. He looked back at Sully, as he dragged his wounded leg along, and decided he would not make it through the night without some help.

A cold finger of dread and terror started slipping down his back. Dylan worried that he had taken his little party out of one kind of hell and brought them into another.

Just ahead he saw a black shape appear and vanish in fog mists. "Did you see that?" Dylan turned around to see Taylor had her head down, blindly following his footprints. She didn't answer.

Staring intently into the fog, Dylan again saw a black figure just on the edges of his vision. This time it looked like a dog. *Maybe it was a black lab playing in the path ahead of them.*

"Here boy," Dylan called. He tried to sound friendly and playful. If there was a dog up here, the camp wasn't far. "Sully, do you see that dog?"

Dylan turned to see the reaction of Sully and Taylor to the dog. Their heads were down, and they seemed barely conscious. *I need to keep them moving.*

Just on the very edges of his vision and through the darkening fog, Dylan could make it out. It was a black lab! A big clumsy puppy! *Damn. That looks nearly exactly like my old dog. It looks like my sweet Bergen!*

"Here Bergen," Dylan called. The puppy reacted strongly to the call but remained on the edges of his vision. *I'm imagining this. Bergen died years ago, shot by a killer in Seward.*

"Bergen! Come here boy!" Dylan called. The puppy wiggled with love and excitement, but kept its distance. As Dylan followed the puppy, he realized he was leading the party downhill and into some woods. It didn't feel right, but if the puppy belonged to the miners, it would likely lead Dylan and the others to the camp.

Sully's progress became so slow, it was nearly that of a crawl. Taylor matched his pace and still the puppy stayed with them, but just on the edge of visual range.

From somewhere nearby, Dylan heard the unmistakable sound of an ax hitting some firewood. Moments later, he could smell smoke and bacon cooking. The puppy joyfully ran in the direction of the sound.

"Hello!" Dylan called with all his reserves.

From out of the fog a Paul Bunyan-like tall man appeared. "Where the hell did you come from?"

"We're lost," Dylan said. "We need help."

"Let's get you inside and see what you need," Paul Bunyan said.

"We wouldn't have made it here if your dog hadn't led us," Dylan said. "What's his name?"

"Dog? There are no dogs around here."

Dylan thought about the dog he had followed to the camp and wondered if he'd imagined it. *It seemed so real—as if my long, dead pup was leading me to safety.*

Dylan knew that on the onset of hypothermia, people could imagine things. *That's what it had to be,* he thought. Still, Dylan found himself looking for the dog.

64

THE CRASH OF THE VACUUM bottle against the hardwood floor silenced the men. Jane had to get out and fast. She noticed a small, ugly folk-art painting of an Eskimo hanging above the place where the bottle fell. She ripped it off the wall crumbling some of the artsy rotten frame, and laid it near the bottle. A scrambling noise from the sitting room told her some men were rushing toward the office.

Just as she slipped into her office and closed the door, she heard the men crash into Bolton's office. Silently, she turned the deadbolt so they wouldn't be able to open the door without finding the key.

Her heart hammering, she leaned against the door to see if she could hear what the men planned to do next.

For a few moments, she heard sounds of the men searching Bolton's office, then she heard Bolton's big voice, "Here's the problem, this picture fell down."

The soft British accent of the Nigerian man said, "How could a picture just fall? Someone is in here."

She shivered uncontrollably as someone rattled the door.

The Saudi man spoke, "It can happen. Look, that frame was old and wormy. I was in my fourth wife's room with her and a picture with a bad frame fell from the wall."

"Fourth wife? Four of them? Why do you guys do that?" the South American accented voice said.

"You should try it. My newest wife is already pregnant, and she's only 14," the Saudi bragged.

"Fourteen? Who would want to marry someone that young? What do you talk about, Barbies?" the British accent conveyed astonishment.

The men on the other side of the door disgusted Jane. She heard some joking about sex as the men seemed to relax and leave the room.

Looking around her office, she grabbed a decorative tea set from a display case. It had come with the other office decorations.

She feigned confidence as she walked out to the foyer where the guard was putting the last piece of the stinky pizza into his mouth, "Hey, that hit the spot. It took you long enough to find dirty dishes."

As Jane placed the tea set on the cart she looked around for items to take down to the kitchen. "I'll get canned if I leave anything up here. Did you notice any other dishes around here?"

"No," he wiped his face on a white napkin and tossed it onto the cart.

Jane swiped her card in the service elevator card reader knowing that it would not call the elevator. A red light blinked indicating the card didn't work.

The guard looked at her significantly. She swiped the card again and the red light flashed. "Sometimes this card needs to be swiped three times before it works," Jane explained. She didn't know what she would do if it didn't work again.

Just then a rumbling sound came from the service elevator. It sounded like the car was approaching. The guard moved away deciding to patrol the corridor. Just as he turned the corner, she rushed around to the left and called the guest elevator using her card. It opened nearly instantly. She rushed back, wheeled the cart in and pressed the down button.

The elevator doors whispered closed and the car started smoothly. Jane wondered if Cheek would figure out a way to put her in jail. Cheek was right about Greg Yeung and the environment.

Could the Colonel really have evidence that Mr. Yeung is an atheist? It didn't make sense. Yeung attended church often and talked like a religious person. Bolton must have been lying to convince his business partners. *What else was he lying about?*

Jane had worked hard to get this far in the Super PAC world. She wondered what would happen when this 501(c)(4) self-destructed. Her career would end. She'd have to get a job someplace unassociated with non-profits and high finance.

It didn't seem fair. Jane wished she knew if she should fight to keep her head above water, or flee to the lower 48 and get a job at Starbucks.

Riding downward in the silent elevator, Jane gazed at her reflection in the gilt-framed beveled mirrors. She looked like a kid playing dress up. *I need to talk to someone.*

65

JANE EXITED THE ELEVATOR pushing the chrome cart with the remains of the pizza and knew she had to talk to Shannon Davis. Jane had met Shannon her first week in Alaska and was surprised to find they were from the same small town in Texas, Nacogdoches.

They'd never met in town, but Shannon's dad was Jane's track coach and her mom taught music at the small state college in town. Shannon's parents were both active in the African Methodist Church and respected by the community. During the last several months Jane and Shannon developed a trusting relationship built on their common hometown.

Like nearly everyone else in the small East Texas town, Jane's father worked at the huge poultry processing company, and even though he was in the administrative office, he still came home every day smelling like chicken guts. The whole town stunk, and Jane wanted to live a different life. Now she found herself craving the normalcy of smelling chicken everywhere.

That seemed like a long time ago. Now Jane ran a big non-profit, and Shannon was marketing director at the Netsvetov Hotel. They were each far from Nacogdoches, but it served as a shared bond.

Jane knew that Shannon often came to work very early to manage communications with the east coast, but decided to see if she was still in her office at this late hour. The office was dark and locked up, but a blue-blazered woman with tattoos told Jane to try the bar.

The hotel bar never ceased to fascinate Jane. Unlike the smoky bars in East Texas, the Netsvetov exuded clean and class. It was truly a place where the high rollers of Alaska could conduct casual business. Visitors often described the bar as cozy, probably due to the several fireplaces burning real wood, but it also projected a chic international vibe. The service and drinks were excellent, so it seemed the perfect place for Jane to unwind with someone from her own hometown.

Jane noticed that Shannon, looking tall and confident, was talking to several hotel employees near the bar. Shannon dressed in a power business suit much different from Jane's.

As a slender black woman, Shannon felt she looked best in clothes with long lines. She wore a white, sheer-back wool blazer with matching wide-legged pants, a peach-colored silk blouse and cream-colored open toed sandals. Jane loved Shannon's style, but it seemed unapproachable to her. Jane thought of her own angular body and wished she looked more curvy.

Jane waited for Shannon to finish her conversation. Soon the others were gone, and Jane exchanged a long hug with Shannon and took a stool next to hers.

"What are you drinking?" asked Jane.

"Club soda, want some?" smiled Shannon.

"I need something stronger, and I need to ask your advice," said Jane.

The two agreed to have something with cranberry juice and vodka.

Jane told Shannon about the conversations she heard upstairs and about the threats and harassment that Cheek had made against her. Running a top Super Pac was her dream, but she might even go to jail because Cheek says she's part of the fraud.

Shannon listened as Jane unloaded her guilt about supporting Greg Yeung's election bid since he was so much against environmental regulations. Jane fretted about her part in hastening climate change in Alaska.

"Honey, y'all got some bad boys surrounding you," Shannon said, dropping her ivy league accent and sounding like a Texas girl from Nacogdoches. Jane loved the way they would both talk in deeper accents when they were together. "Maybe you should swear off men for a time."

"Good idea, Shannon. I'm really scared," admitted Jane. "I could lose this job and any future work in the field I love."

"Poo. Don't let that chickenshit Cheek scare you. If he could have busted you for your so-called 'crime', he would have done it long ago. I'm no lawyer, but he doesn't have anything on you."

"But what about his proof, showing part of an email I sent the Colonel?" Jane asked.

"Girl, you send copies of all your emails to me. They can't get to them on my hotel server. It's tighter than a clam's ass at high tide. I think you should worry more about what your candidate will do to the environment if he wins. You need to be on the right side in this fight. I don't like you on the side of big oil." Shannon stirred her drink.

"This is how our country works. It's an adversarial model. Both sides offer their best to voters, and then democracy works. Each side deserves to tell their story."

Shannon could see she shouldn't argue with Jane on this point, so she checked her phone. "Jane, I gotta go. But I want you to talk to me when you get some good news about this shit."

Jane smiled, "I want to do that, too."

66

PAUL BUNYAN TURNED out to be a massive Minnesotan named John Jolly. John worked a claim with a friendly group of three Hawaiian men. They met in the early spring and worked the claim all summer until the harsh fall weather drove all of them back to the beaches of Kauai. It was a spare, subsistence lifestyle that Dylan understood.

The first thing Dylan wanted, after drinking lots of water, was to call authorities about the attempts on their lives. However, John explained that their sat phone was in pieces on his workbench, a victim of an errant sluice box, but he would have his tech expert work on re-assembling it as soon as he finished repairing the sluice box.

John had hurt his wrist a week earlier, which took him off his mining duties and allowed him to devote a lot of time to his new guests. He saw to the rest and rehabilitation of the ARCS party, while the other miners continued working their claim. Besides how dirty they all were, John was surprised by Sully's ability to consume food. Sully doggedly worked his way through the expired MREs they had lying around.

What surprised Sully was that no one liked the jambalaya meals. John Jolly considered them inedible. John's cooking wasn't much better. After the ARCS team figured out that the only decent meal John could fix was pancakes, Dylan took over the cooking for a few meals.

Taylor attended to Sully's wounded leg, and all of them caught up on showers, laundry and sleep, but none felt at ease. Well-organized deadly killers, one of whom had been dressed like an Alaska State Trooper, had hunted them.

John walked into the insulated mess tent and held up a newly repaired satellite phone. He wanted to know who he should call to report their rescue, and Ethan's and Zhang's death. Dylan told him not to call anyone until he and the others could discuss the matter. "Also, you should know that our phone is a WorldSat model that works great in Hawaii where we bought it, but sucks this far north."

After John left the tent, Dylan, Sully and Taylor munched on cold pancakes and hot coffee while they figured out their next steps.

"We should charge our phones and call ARCS and tell them about those shooters. The ARCS staff might be in danger," Taylor said.

"There are no cell towers out here, so your phones can't catch a signal," said Dylan.

"And we might be in danger when we return to ARCS. Someone wants us dead," Sully rolled up some butter and jam in a rubbery cold pancake as he fiddled with his newly recharged data recorder. "OK. There's no phone signals up here, and anyway my phone is broken."

"And we need to inform Dave of Ethan's and Zhang's death. Maybe Zhang was killed, too," Dylan said.

"We should be careful. It could be that all of those bad guys were troopers," said Taylor. "Maybe there's corruption within the law enforcement community. Maybe we shouldn't call the state police."

"All the troopers I've ever met were solid cops. Maybe that guy who blew himself up wasn't a real trooper," Dylan sipped his coffee. "It could just be something funny going on at ARCS, but you are right. We need to call Dave White and tell him what's going on."

"Not only that, we need to figure out why so much of the data in this nuclear magnetic resonance spectrometer is different from what all of us at ARCS have on our computers. Maybe there's someone or something that is corrupting the data on the servers at ARCS." Sully licked some jam running down his wrist. "I found some other weird data on this spectrometer just now."

"What do you mean *other weird data*?" Taylor scooted next to Sully and looked at the display of the data recorder.

"This seems to indicate that *the Scrubber* was not promoting bacterial growth and in turn not scrubbing CO_2 from the exhaust gases," Sully pointed at a line of data.

"So your new data is supporting our cave conclusions: no bacteria and consequently high CO_2 levels. But your membrane technology is supposed to be superior to my lab test tube in that it provided more favorable conditions for bacterial growth and CO_2 reduction." Dylan stood behind the other two scientists and stared at some incomprehensible numbers and symbols on the screen.

"My laptop has all the software and data I would need to analyze this new spectrophotometric data. Let's get this data back to ARCS for analysis." Sully stood up and winced.

"That leg still causing you pain?" asked Dylan.

"I'm not feeling comfortable about going back to ARCS until Dave can look into those shooters," Taylor said. "But we do need to figure out whether the bacteria grew during the last run and your leg warrants a professional assessment."

"When I first saw the Scrubber in my daughter's office, she showed me a notebook with all the preliminary test data and design specifications, but sadly I didn't have time to look at it. Apparently, Professor White had presented her with a copy of some of the early research to encourage her participation in the membrane project. Anja promised to send the report to my home in Seward," said Dylan.

"If I had that report, I could complete my analysis and then probably figure out why there's a discrepancy between what's on this spectrometer and the ARCS data," said Sully.

"We should go to Seward and examine that notebook," said Dylan. Dylan felt a surge of excitement just thinking about going home. He loved this wild part of Alaska, but this world of scientists, politics, ghost dogs and killers filled him with a profound uneasiness.

He wondered if the shooters were really after him and not the ARCS team. Maybe all this was connected to something in his life. He thought of Suzie, far away in Africa. Anja was safe in her Anchorage office with all of its security. Probably the best way to feel safe was to resolve the mystery of the conflicting data.

"Let's go to Seward and examine that notebook," repeated Dylan. "We'll call Dave and tell him what we know. We'll explain that we're going to check out *the Scrubber*'s preliminary reports. He needs to know if something's funny with *the Scrubber*."

"He also needs that coal. And maybe you shouldn't mention that we're going to Seward if you are talking on an unsecure line. Anyone who can hack ARCS servers can listen in on a satellite call," Sully said.

"Seward?" John entered the room. "You can't get there from here. Our supply plane doesn't come in for another day or two, and then it goes to Fairbanks. You're stuck eating my pancakes until then."

DYLAN NEEDED a break. He mentally considered the implications of *the Scrubber*'s preliminary data as he walked the edge of the woods surrounding the mining camp. He felt a nagging sensation tagging along behind him as he followed the slightly trampled snowy paths that might one day become a road. *Is there something I'm forgetting?* He glanced down at his watch. It was nearly 2 o'clock. *Shit!*

Dylan stormed into the mess tent. "I need to make a phone call!"

Sitting in the corner sipping coffee, John nodded towards the sat phone in the corner. "Good luck. Like I said the northern latitudes aren't very sat phone friendly." John studied Dylan. "What's the matter? Forget to cancel the daily paper before you left home? I hate when that happens."

Dylan paused to ponder and then ignored John's question. "Today is Governor Bates-Hardy's speech at the Oslo Arctic Climate Conference. In fact, the governor's talk is supposed to begin at 2 p.m., that's in 15 minutes!"

Dylan checked the *contacts* on his cell phone to find the governor's private phone number.

Using the camp sat phone, Dylan hoped for a signal then tried the number for the governor's cell. No answer. Next he dialed the governor's office. One ring . . . two rings . . . three rings . . . "Hello," began the recorded message. *Where else can I call?* wondered Dylan as he ended the call.

"What's up, Dylan?" Sully was standing in the mess tent doorway.

"Sully, we've been single-mindedly focused on the test results and what to tell Dave. Do you remember what else is happening today?"

Sully looked blankly at the mess tent table. "It's pancake day?" Then realization hit Sully's face like the slap of John's rubbery pancakes hitting a plate. "Shit!"

"Shit is right! The governor's speech starts in twelve minutes! If she announces *the Scrubber* and it turns out not to scrub, it could ruin her! There was no answer at either the governor's cell or the governor's office."

"But a satellite phone line is not secure!"

Dylan considered the security issues. "Well, I'll just obliquely advise the governor that we've generated some new data. The governor must not act as if we have locked in one specific conclusion. We just need a little wiggle room until we can sort this out."

Dylan re-dialed the governor's office number.

67

THE GOVERNOR'S NEW student intern, Alicia, a blond square-jawed, former state champion softball pitcher from Eagle River, eyed the telephone beeping on the receptionist's desk. The governor, the receptionist and most of the staff were downstairs in the ballroom—the site of the governor's televised speech to the Oslo Arctic Climate Conference. The speech would be simulcast to all the delegates in Norway and elsewhere to North American academics and supporters of the governor.

Within another minute, the phone rang again. Alicia felt as if the call was urgent, maybe even desperate. Her instructions had been very clear, *remain available for any last minute errands or requests.* Answering the phone was not on her list, but the phone repeatedly would ring, pause, and ring.

Alicia picked up the phone. It was an advisor to the governor calling on a satellite phone from the wilderness. He had a terribly urgent message that the governor needed before she started her speech.

Alicia scribbled the message down on a post-it and assured the caller she would do her best to get the note to Governor Leola Bates-Hardy before she delivered her speech.

Alicia had never talked to an advisor to the governor, nor anyone on a satellite phone, and especially someone who was so well informed they could stop a speech by Governor Leola Bates-Hardy! *I must get this message through to the governor!* thought Alicia.

THE HALLWAYS were crowded as Alicia skirted a video team, sound engineers and huge ropes of black cables taped to the floor. She rounded the corner and navigated past temporary scaffolding.

She started to second-guess herself about answering the phone, but she could sense the call's importance. *And I was right*, she thought. The caller was the governor's technical advisor.

ALL DYLAN could do now was wait while Alicia attempted to locate the governor before the sat phone lost the signal. An open area of the mess tent provided little room for pacing back and forth, but Dylan walked where he could. He hoped the intern would reach Governor Bates-Hardy before the governor's speech began in minutes. And, he hoped the governor would quickly return the call from Dylan.

Dylan spied a laptop on one of the mess tent tables. "John, I'm waiting on a return call, but after that can I send an email from here?"

"Yes, but you need to route all email through the sat phone." John Jolly made the gesture indicating it would cost money.

"I'll pay for the time." In his mind, Dylan began composing an email to Professor Dave White. Dylan's message summarized the chase and Ethan's death and indicated to White that Taylor, Sully and Dylan were heading to Seward to recover a lab notebook that should help explain the spike in energy use by *the Scrubber*. Not knowing whether White frequently checked his email, Dylan planned to address the email to the ARCS secure mail site.

Come on, Hardy. Call me! Dylan mentally practiced his summary to the governor. *The Scrubber's data has been falsified. Don't report that the ARCS's data validates the performance claims.* Hardy must not tie her reputation to a claim that appears to be based on falsified data.

Dylan noticed he had started oval breathing. Presenting falsified data at the Oslo Arctic Climate Conference was a double-pronged political disaster. The governor's credibility would be brought into question at a critical time during the election process. And, given Anja's support, Mountain Club's credibility would also be tarnished. The whole scrubber project will look like a political or financial sideshow rather than an environmental priority.

Eight minutes until the scheduled beginning of the governor's speech. Dylan cycled his breathing. *What was the hold up?* He suspected that the distance between the receptionist's desk and the Governor's Mansion Ballroom could be quickly traversed. Dylan hoped the presentations were running a little late. The governor's address was scheduled just before the lunch break and following several remotely broadcast sessions.

GOVERNOR LEOLA Hardy-Bates glanced up from her notes feeling both the excitement and the anxiety of the moment. Seeing the endless cow images presented by the current speaker invoked a smile, lifted her mood and generated some notes for Leola's presentation.

If only Anja Hart would have accepted her invitation to stand by her during her presentation, thought Leola. Anja's appearance and the implied connections with both Dylan and Mountain Club would have been a wonderful coup. But, unexpectedly, Anja had declined the invitation.

The governor set her notes aside and deliberately relaxed.

ENTERING FROM THE REAR of the ballroom, Alicia was struck by the huge projected image of a cow and the several pie charts on easels that framed the speaker and the podium. A laser pointer dot jumped from bar graphs to pie charts as the speaker emphasized take-home messages.

Alicia remembered seeing this guy on the agenda. The speaker, Mike somebody, was proposing to supplement cattle feed with methane-consuming bacteria. As these bacteria grew in the cow's stomach, methane gas production by the cow would be dramatically reduced. Fewer cow farts meant less methane dumped into the environment. The governor's conference team had jokingly referred to this speaker as "Methane Mike".

Alicia spotted Governor Bates-Hardy on the far side of the podium. Alicia could see that the news media and audience members were patiently waiting for the governor's address. Alicia rounded the last row of seats and headed directly toward the governor as Methane Mike stepped down to mild applause from both the ballroom and the auditorium in Oslo. Alicia watched the live video feed on the center screen as the Oslo moderator, framed by the Norwegian and EU flags, began the governor's introduction.

As Alicia approached the edge of the stage the first of three summary slides appeared on the big presentation screen. At that same instant the governor's senior political advisor materialized in front of her blocking her path. "I have an urgent message for the governor," Alicia informed Cheek.

Harold Cheek towered above the student intern, his stern expression and body odor making her step back. He was not pleased. "Have you missed the fact that the MC is currently introducing the governor? This is not the proper time to disturb Governor Bates-Hardy."

"But the governor's technical advisor told me that his information influences the conclusions of the governor's presentation. He's in a remote mining camp talking on a sat phone!"

"I'll place a return call to the technical advisor after the governor speaks," said Cheek.

As Alicia began her response, Cheek interrupted, "Thanks for your concern, please return to the office."

As she retreated, Cheek crumpled the message Alicia had given him. *No distractions! That asshole mountain man can wait until after Hardy's speech!*

68

BOLTON SAT on a leather bench beside a huge fern as he waited for Greg Yeung to come out of the locker room. With large metal abstract art, clean locker rooms and fluffy towels, Bolton felt the Netsvetov Hotel executive gym was as nice as any he'd ever used. He was thinking that if Asians didn't dawdle in the showers so long, he and Greg could be back to work on Yeung's campaign.

Pulling his phone out of his pocket, he connected to the robust hotel wifi and began scanning his usual media sources to check on his investments and scan news stories related to his businesses.

It wasn't until he got to news stories related to the energy industry that he learned Governor Bates-Hardy was addressing the Oslo Arctic Climate Conference live at that moment.

He checked the hallway where he expected Yeung to appear, but it was empty. *The little bastard was probably still in the shower. What's he doing? Trying to get the brown off?*

He clicked on the link to the governor's speech and watched the video, which was being live-streamed to conference attendees in the huge Oslo Conference Center, but originating from the Governor's Mansion in Anchorage.

Bolton wondered what the governor could possibly say that would interest a bunch of Norwegians and other gloomy climate geeks. *Will she announce the opening of new landfills to lower the cost of garbage dumping?*

Just then, Greg Yeung sat down beside Bolton smelling of some kind of sport-themed soap, shampoo and deodorant. "What are you watching so intently?" Yeung asked.

"Check this out. It's an address to the Oslo Arctic Climate Conference. The governor of America's largest state bills it as an important announcement.

"What could she announce that the climate geeks would care about?" Yeung asked as he leaned closer to Bolton to see the images on his phone and bringing a stronger sports shampoo odor to Bolton.

"She'll probably just welcome the conference attendees and list some of Alaska's concerns that she thinks need to be addressed. This speech isn't watched by anyone but climate freaks in Alaska, so it probably doesn't matter what she says," Bolton said. "Maybe she'll discuss some local issues. Alaska's economy is in a slump. Unemployment is up. Electricity rates are going through the roof. Maybe she'll announce the licensing of a new natural gas-fueled power plant. That would really help Alaskans. I would sell Alaska natural gas from my future wells."

TALL AND CONFIDENT, Leola Bates-Hardy strode to the podium accompanied by enthusiastic applause by the friendly live audience in the ballroom. She smiled facing the cameras and lights.

"Did we discover enough about cow farts today?" The audience laughed quietly. "I see a lot of people shifting in their seats. Maybe we should put that methane reduction bacteria in the bean dip over there on the snack table." More laughter.

"I've come today to announce a discovery by Alaska scientists along the same lines as cow farts. This discovery should change our environment, not only here in the great state of Alaska, but perhaps even the world. But first I want to thank . . ."

The governor went on to thank the conference organizers and shout out names of dignitaries. While she did this, suspense and curiosity began to build in the audience. *What discovery could change the world?*

The room darkened, a screen lowered behind Leola and a projector came on showing an image of a gorgeous brilliant-cut black gem against a blue silk back-ground.

"If I asked an expert what gem this is, one guess might be black jasper, but look at the translucent corners of the stone. Black jasper is opaque." The next slide zoomed in to show a close up of the gem revealing sharp light shining through a beautifully cut edge. "It's not black jasper."

"The next guess might be black agate, but we can find no evidence of any banding." A new slide appeared viewing the stone from another angle revealing a pure blackness that could be some kind of black glass.

"So most gemologists, based on these photos, would guess a very pure onyx." The governor held up a black, grapefruit-sized stone—the very stone in the slides. A spotlight from above ignited the facets in sharp, reflected light that dazzled the audience as Leola turned the stone in her hand sending scattered light into the eyes of the audience.

"This is not onyx. This is a sample of coal from a new discovery a few hundred miles from here." The audience stared in rapt attention. "Not only is it coal, but it's so pure, it could be burned without releasing any of the heavy metals normally associated with coal combustion."

"You are thinking, 'Not more coal! It only throws more carbon dioxide into the air.' And how many speakers today have mentioned carbon dioxide's environmental damage to the earth and especially Alaska?"

AT ARCS, CONRAD had set up his large monitor in the crowded break room so the staff could witness the announcement. All work had stopped as the staff gazed at the computer screen, waiting for the governor to announce their progress on *the Scrubber*.

Sequestered in his stuffy office, Dave White watched his tablet and listened with a rising sense of dread in his chest as he watched the governor move to the next slide. *I need more time!*

The governor's next slide revealed an artistic photograph of a set of lab glassware. A bubbling round-bottomed flask was highlighted and the rest of the glassware appeared in shadow. "An Alaskan, Dylan Baker developed and patented a process you need to know more about."

Leola advanced the slide to show a close up of the flask. "Inside this flask, a very hungry bacterial culture eats carbon dioxide. It craves carbon dioxide. It wants more and more and more." The governor spoke enthusiastically.

The next slide highlighted another part of the glassware that looked like it might be part of some distillation machinery. "Once the carbon dioxide is consumed by the bacteria, a waste product, flammable alcohol is produced and removed using this extraction apparatus."

The next slide highlighted a clear liquid inside a collection beaker. "Once the alcohol is removed, it can be used as a high-energy gasoline additive." As the implications of the presentation hit the audience, a powerful hush fell over the crowd.

As the governor went on with her speech, it seemed her audience's spirits lifted to unimaginable heights. When she explained the exhaust stack filter the ARCS team had designed, the audience went into rapture. To have a clean burning energy source that produced combustible alcohol was far beyond their dreams.

This kind of resource discovery coupled with a huge technological breakthrough, could provide an enormous boost to Alaska, reduce greenhouse gases and provide a soft landing for the fossil fuel based economy until renewable energy technologies matured.

BOLTON'S PHONE made a *ping* sound every time an email came into his IN box. The *pings* were so frequent; it shut off the audio from the Oslo Arctic Climate Conference.

AT ARCS, A CHEER RANG out in the break room as Dave White sat in anguish in his small, dark office. *More time! Just a little more time and we'll have it.*

THE NEXT SLIDE showed a photograph of Dylan Baker getting into Flapjack's Beechcraft Bonanza.

"Over 30 years ago, Dylan Baker patented the process for using bacteria to remove greenhouse gases from the atmosphere. Since then, he's observed the amazing progress Dr. Dave White has made in improving the process and scaling it up for industrial use. Dylan Baker is a true Alaska hero. Due to his discovery, Alaska's Permanent Fund is projected to shift to the black. No more income taxes!"

69

HAROLD CHEEK FOUGHT hard to control his anger as he ducked through the door and into a hallway. Striding down the hallway at a brisk pace, Cheek rounded the corner and stepped outside under the overhang where the smokers hang out. Of course it was raining. Nothing was going right. He punched in the number for Mountain Club and prepared to go through all the phone menus in order to reach a live person.

Cheek thought about Anja Hart, an insatiable workaholic, she would probably be at her office. If Anja had been unable to attend the Oslo Arctic Climate Conference in person, she was probably at this very minute watching the on-line web-cast version. In fact, she was likely wondering why she hadn't been better informed.

The grounds outside the Anchorage governor's mansion blurred as Cheek watched the big Alaskan rain drops splash in a puddle. He mentally reviewed the events of the past few hours. He could not believe Dylan Baker! First, Dylan calls two minutes before the governor's speech, *insisting* on speaking with the governor.

Then, Dylan goes on to explain that he has some new information that contradicts the positive performance data the governor presented. Was Dylan trying to destroy everything? What could Dylan hope to accomplish by calling the governor at the last minute?

Cheek had not been happy about sending Dylan to ARCS. But, the governor's logic was sound. As father of Anja Hart, Dylan Baker might get The Mountain

Club to cough up some needed campaign money and an endorsement.

Right now, the issue on the table was damage control. Dylan was a smart man. But if he were planning to go up against the governor's staff and re-election team he would need some powerful help.

Cheek was pleased that the governor had secretly recorded Dylan's comments during their face-to-face meeting. At least there would be some insurance if, for some reason, Dylan changed his mind about supporting the governor.

"Mr. Cheek?" the voice on the line reigned in Cheek's thoughts. "Anja Hart. I'm rather surprised that you called."

"Hello Ms. Hart." Cheek could sense the some tension in Anja's voice. "Do you have a minute?"

"I would have thought that everyone would be celebrating. Great presentation! I bet the re-election team is energized with Bates-Hardy getting back some momentum."

Is she pissed? wondered Cheek. Mountain Club's lead fundraiser and the public had learned about of the latest scrubber data at the same time. Now there was no opportunity for Mountain Club to prepare appropriate responses and insure the best political spin for the announcement.

"I'm sorry Ms. Hart. We were unable to give you an advance copy of the speech."

"Sorry for what? Neglecting to keep your allies current? Allowing us to waste time preparing statements for unlikely scenarios?"

"Again, I'm sorry. You know the race for governor is too close to call. And you know how candidate Yeung and oil-advocate Bolton are not missing a beat when it comes to capitalizing on anything that may reflect poorly on Governor Bates-Hardy." Cheek needed to get off of the defensive. "Have you spoken with Dylan?"

"No. The governor had asked Dylan to visit ARCS and observe *the Scrubber*'s performance. Why do you ask?"

Cheek exhaled a deep breath; he had reached Anja before Dylan. "I suspect you'll be getting a call from Dylan sometime soon."

"Good. I was concerned that Hardy might have con-scripted Dylan onto the re-election team. I'm pleased that she hasn't put that kind of pressure on Dylan."

"Leona Bates-Hardy is an honorable person. It's Dylan I'm concerned about."

"Dylan? Why Dylan?" said Anja.

"Look, Dylan meets with Dr. White and the ARCS team. He's bubbling with enthusiasm. It makes sense. The Scrubber is based on his early experiments. Then this afternoon Dylan did a one-eighty and he began telling people that the Scrubber's performance data was flawed and maybe even faked. Dylan even tried to discuss his new information with the governor moments before she gave her presentation!"

"That's absurd. Dylan clearly understands the science behind *the Scrubber*'s performance. It's difficult to imagine him flip-flopping in his support for this project."

"I sent you the link. You saw the video of Dylan's assessment of *the Scrubber*. And you heard him concur with the conclusions based on the original data. Well he called me today, minutes before the governor's message, and tried to convince me to cancel the presen-tation," said Cheek.

"On what basis?" Anja said.

"He wasn't specific. He said he had identified some inappropriate assumptions and flaws in the data."

"Really? Dylan doesn't strike me as someone who would change his position without good reason and sup-porting data." Anja sounded puzzled.

"Bolton got to him," Cheek said. I might guess Bolton is like a father-in-law to him. "I've received reports that

Dylan has talked with Bolton. Don't you think the timing of Dylan's chat with Bolton, and then Dylan's one-eighty on *the Scrubber*'s performance seem like a suspicious coincidence?"

Anja paused to collect her thoughts. "Yes, he has spoken recently with my biological grandfather. I don't know what was discussed. Bolton and I don't talk. But I do know that Dylan and Bolton are not on good terms. The conversation would have been very tense."

Cheek continued, "Maybe Dylan has ulterior motives. Maybe there are some financial opportunities for Dylan here that neither you nor I have anticipated."

"Mr. Cheek. Is Dylan still at ARCS, or do you have him squirreled away in that mansion somewhere?"

"Dylan is unavailable for phone or email communication. When he called the governor, he was in some remote mining camp talking on a satellite phone," said Cheek.

"What the hell?" Anja burst out.

She knows more than she's letting on. I need to get more direct, thought Cheek. "Look Ms. Hart, this call is about Dylan. Today's call is a courtesy. This is my attempt to keep the lines of communication between our groups open. But I want you to know that Governor Bates-Hardy places a very high value on integrity.

"If Dylan abuses the faith placed in him, he'll fall, and he'll fall hard." Cheek paused for emphasis. "I expect you to repay the courtesy of this call with a return call when you hear from Dylan. Have a good evening, Ms. Hart." Cheek hung up.

HAROLD CHEEK'S DIRECT, unfiltered comments surprised Anja. She considered how much this call troubled her. Anja also decided her connection with Dylan needed some attention. She saw advantages to maintaining that relationship.

A tone pinged from her desk. Anja looked down at the touch-pad surface and noticed message indicted a call was waiting for her. It was Dylan.

70

AS IT LANDED, the green and white Cessna 172 bounced down the slushy runway on its fat tundra tires, splashing muddy snow everywhere. All work at the mining camp stopped as the miners and ARCS team approached the place where the taxiing plane would stop.

"Hey, it's Gator," called out one of the Hawaiian miners.

As the propeller hiccupped and stopped, sudden silence rushed into the camp. The pilot's door opened and a short, stocky pilot stepped out wearing a University of Florida Gators baseball cap and a worn leather sheepskin jacket with its bottom snaps unfastened.

As the figure approached the group, Dylan could see that the pilot was a fit-looking young woman sporting a bobbing brown pony tail who strode confidently up to the miners while holding out a box of fresh donuts.

"I gotta pee. You gatorbaits had better have toilet paper in there this time," she announced as she handed off the donuts and walked past the genuine smiles of the men towards the slightly leaning outhouse nearby. The miners quickly went to the cargo hatch and unloaded the items they were expecting while munching on their treats.

"She's pregnant. Has to pee all the time," John Jolly explained to Dylan. Dylan nodded as if he understood.

"Think she'll take us to Seward?" Sully asked, as if pregnant women couldn't go that far.

"If she has time in her schedule, she'll do it. But her Cessna's capacity is four passengers and fuel. She needs to take our ore samples plus whatever luggage you have." John Jolly looked at Sully's generous waist. "So every pound of cargo means less fuel she can carry."

"Perfect," said Taylor. "Then she can stop for fuel and potty breaks along the way."

A few hours later, the little plane's cramped interior was packed with the ARCS team, ore samples, Sully's data recorder and a few other items. Gator had explained to Dylan that their WorldSat phone would probably catch a satellite signal once they were at 5,000 feet.

As she went through the preflight check off, Gator gave her *Welcome Aboard* speech.

She pointed to Sully. "What's your name?"

"I'm Sully."

"OK. Call me Gator. You're in the front seat because you're fat. I need to balance the plane, or I'd put you in the back. You look like you're going to try and fiddle with my controls." She faced him. "Are you?"

"No Ma'am," Sully said.

"No Gator," she corrected. She looked at Dylan squeezed into the backseat next to Taylor. "I don't want you or your pretty little wife barfing in this plane. Get it?"

"Yes Gator," the back seat residents said together. Neither decided to correct Gator when she was on a roll.

Gator nodded and continued the preflight. Soon she was settled in the pilot's seat, quickly slapped Sully's hand when he reached toward a knob, then turned to make sure no one was fooling around in the back seat.

She opened her door and bellowed, "Clear the prop!" The yell was so loud that it made everyone jump. "I used to be a gym teacher; that's where I got my big voice," Gator explained.

A few minutes later the little plane was airborne in a clear blue sky. Denali, all white in the sparkling sunshine captured everyone's attention as the green and white plane struggled to gain altitude.

"I want to go over our itinerary again," Gator began. "We're flying into Fairbanks because I am dropping off the ore samples there. We'll be on the ground for about an hour. Then we're flying to Anchorage because I want to Facetime my daughter, get fuel and pee. Finally, we'll head to Seward."

"That sounds fine with me," Taylor said. "Tell us about your daughter."

Gator softened, and talked non-stop for the next 20 minutes about her wonderful four-year old. Dylan, thinking about his daughter, had to interrupt her to tell her he was going to try a call with their sat phone.

"What? Am I stopping you?" Gator barked. "Just plug your headset into the phone so you don't hear me."

Dylan turned on the phone and instantly got a strong signal. He called the governor's office number and left a message explaining that someone tried to kill them and that they found serious problems with *the Scrubber*. He asked the governor to call him back.

"Are you sure you want to talk to the governor? She's the boss of the state troopers. Maybe she sent the bad guys?." Taylor spoke in a low voice.

Dylan explained to her that he trusted the governor, and that he thought the trooper was probably not a real law enforcement officer. Then he said he needed to call Anja before she called Professor Dave White at ARCS.

He punched in Anja's work number and spoke to an assistant. She asked him to wait. His sense of anticipation grew with each minute he was kept on hold. Finally his daughter's voice came through. "Dylan? I was told you are calling from a small airplane."

The sound of her voice made him happy. At that moment, he ached to talk to Suzie who was somewhere in a wild part of Africa. "Anja! We have a problem with *the Scrubber*."

Dylan explained the discrepancies of the test results while she listened in silence. Then he told her about

Zhang and Ethan's deaths and how the remaining team members barely escaped with their lives.

When he had finished, Anja nearly shouted, "You could have been killed! This is terrible. We've had some security problems here recently, too. But nothing so overt as you've been facing." The intensity and the honesty of Anja's reaction struck a strong fatherly cord, perhaps it was the beginning of the father-daughter relationship Dylan has been seeking.

Gator passed around a bag full of candy bars. Sully took two Snickers, Taylor grabbed an Almond Joy and Dylan pulled out a Kit Kat. It surprised him how reassuring something normal like opening a candy bar was to him.

"Did you hear me?" he heard Anja's voice. " I asked if it were possible that the ARCS data was correct and your data recorder has errors in it."

Dylan acknowledged that it was possible, but pointed out how the data recorder held the raw data, and it indicated that *the Scrubber*'s performance tailed off dramatically near the end of the experiment indicating the bacteria all died. If it stopped working as the raw data indicated, it was pretty much worthless.

"What are you going to do about it?" Anja asked. "I don't want you to take any unnecessary risks."

Dylan smiled to know that Anja cared about his safety. "We're going to Seward to compare the information in the notebook you sent to me with the data produced by the experiment. We might just figure out what's going on." Dylan munched on his Kit Kat.

"If someone is tampering with *the Scrubber*'s data, who could it be?" Anja mused. "It has to be someone who wants to convince the public that we can do something about carbon emissions when in reality the equipment is ineffective."

IN ANCHORAGE standing in a line at a crowded, noisy coffee shop, Senior Political Advisor Harold Cheek ordered a decaf soy latte with an extra shot and double cream and looked down at a new text message on his phone: *They are on the way to Seward to attempt to prove the Scrubber does not work!*

"What the fuck?" Cheek shook his phone as if to make the message disappear.

A confused barista looked at the order on Cheek's cup, "Sir, you want a soy latte with cream?"

71

LOOKING DEFLATED, candidate Greg Yeung stood up as if mimicking a robot's soulless movements. The sports shampoo odor instantly cleared around Bolton, "What just went down at that Oslo conference? I got to talk to my broker." He left Bolton sitting catatonically staring at his phone.

Bolton stood and wobbled as if about to faint then wandered to the elevator and punched the code to go up to his penthouse apartment.

Once he was alone amongst the cold glitter of his overly-decorated penthouse, Colonel William Bolton lifted his Colt Paterson revolver from its display case. The 5-shot, 1838 gun appeared to have no trigger or trigger-guard. Bolton pulled back the hammer and the trigger popped down. The gun was ready to fire.

Years ago, Bolton had gone through the laborious process of loading the gun. This involved disassembling the device using a screwdriver, small mallet and pliers. Then he loaded each chamber with black powder, .28-caliber ball and a percussion cap. Afterwards he had packed a lard-and-beeswax covering over each chamber. Since the gun had no safety, he had only loaded four of the five chambers. Now he didn't need it to be safe.

He had bought the gun from a seller who claimed a Texas Ranger used it during a famous Indian war, but he couldn't remember which war. He could not imagine loading the gun during a battle. It must have taken nerves of steel.

Right now his nerves were gone. He noticed his hand trembling as he held the gun. His world was gone. With the discovery of the giant seam of pure coal and *the Scrubber*'s method of removing the CO_2 from the combustion of coal, all his dreams had vanished.

That tall bitch, Bates-Hardy would be re-elected and all his drilling rights would be worthless. She would see to that. He would be unable to repay his "loans" to the Super Pac in time to avoid getting caught. He'd spend the rest of his life in jail.

As he held the antique weapon, the long barrel seemed to slowly turn of its own volition towards Bolton's face. This was the way to avoid prison. Being locked up with a bunch of losers and horny queers caused jolts of loud, red, flashing emotions to surge through his body. Never in his life had Bolton ever been afraid. *Am I more afraid of dying or prison?*

Unaccustomed to personal reflection, Bolton found the inner voice disturbing. He noticed himself speaking aloud. At least he thought he was speaking aloud. "I'm not going to prison. I have control over this situation."

His voice sounded weak and hollow in the big office. Adding to the powerful sense of defeat and fear in his body was confusion that the world order was changing. *I'm the one in charge. I'm never afraid. Who is this inside me?*

Bolton noticed the gun continued to slowly turn toward his face as if out of his control. He felt his jaw opening as if he was observing himself from above. He looked down and watched himself positioning the revolver so he could place the barrel in his mouth. An odd thought struck him, *I'm getting pretty thin on top. Maybe I should change my haircut.*

The gun barrel tasted metallic and bitter as it entered his mouth. *I need to angle the shot upwards so I don't shoot off my cheek and survive.* Bolton recalled stories about wimps who couldn't do it right. *What assholes!*

Bolton wondered at his mental clarity at this moment. The Colt paused in its journey through history. The clear Alaska sunlight slanting through the office windows illuminated dust motes suspended in the beams. It seemed he could see each tiny particle as if it were enlarged to the size of a walnut turning slowly in a gravity-free sunbeam. Clear thoughts and energy flowed into him. *I should pull the trigger now. The ball will bounce around inside my skull and shred my brain. No pain. Instant oblivion.*

His finger tightened on the trigger. He remembered that the gun had a nice trigger pull and nearly no recoil. It would be good.

Across the room a dark shadow spread toward him. The dust motes vanished as a cloud moved across the sun. *God damned Alaska. This would never happen in Texas.*

As the darkness moved towards him, the gun dropped into his lap. From somewhere a powerful anger gripped him.

That bitch governor thinks she beat me. Anger started pushing back the fear in his mind. The anger came from a familiar place in him. The hyper-competitive colonel's finger hesitated by the trigger. If he pulled the trigger, he would not win, but if he was arrested, he might not have another chance to control his life.

The anger granted energy to him and pushed back on the dark feelings. He welcomed the familiar emotions as the fury bloomed and swelled in his chest. *That fucking governor! She is fucking with the wrong guy!*

As a storm of righteous hostility flooded his office, the pistol fell to floor. The unguarded trigger caught on his trousers and the gun fired into the wall, filling the room with a bitter cloud of acid smoke. Coughing, Bolton stood. Fired indoors, the sheer amount of smoke was astounding.

Above, a smoke detector chirped then went silent when Bolton fanned it with his jacket. As his anger continued to build, his posture straightened and his big chest filled with power. He felt his muscles tighten and the room shrank to a collection of mere physical objects completely under his control.

He would get that stupid fucking governor. He would not let her have her way with him. There are ways to overcome pure science. Look what the tobacco industry did for years when science proved smoking caused cancer. Look what he and the rest of the energy industry did to claim climate change was an "unproven theory".

Just then his doorbell rang interrupting his internal rant. *Who could that be?*

The gun. The motes. The smoke. Everything suddenly shrank in importance. Bolton was back, and he was pissed. The governor would pay for this. He would get back at Jane and anyone else who didn't do enough to keep him informed and protected. *God damn them!*

At that moment, if anyone had asked Bolton if he was in the act of taking his own life, he would have righteously denied it. His memory of the gun in his mouth had vanished from his mind even as the metallic taste remained on his lips.

Gun in his hand, Bolton strode across the room and nearly ripped open the door. A small brown man in an expensive English suit and thick glasses stood there looking up at Bolton owlishly.

"Krish! What the hell do you want?" Bolton shouted unconsciously waving the huge Colt.

INSIDE THE NOISY CESSNA, Sully held up a Snickers with a bite out of it, "Hey, does anyone want this last candy bar?"

Gator's head whipped around, "Holy shit, there were at least ten of those in there. They are all gone?"

Taylor leaned forward, "Sully, you can't consume so much sugar. If you have a diabetic episode up here, it will take us a while to fly you to a hospital to get you stabilized."

"He's diabetic? Damn, Sully. Knock off the sugar," Gator turned to face Dylan, who was still on the phone with Anja, "Hey Skinny, grab that beef jerky behind your seat."

Gator looked at Sully, "If you eat some protein along with the sugar, it will slow down sugar absorption."

"I know that. These candy bars have nuts in them. That's protein. Now everyone quit picking on me." Sully ripped open the jerky package provided by Taylor and started in on the contents.

Dylan tried to shut out the confusion in the cramped airplane, "Anja, what did you just say? It's noisy in here."

"I asked you who could be trying to kill you guys."

"It could be anyone who wants to sabotage this carbon reduction project," said Dylan.

"It would also need to be someone who has the ability to get a team of shooters on the ground. That takes money and logistical support," pointed out Anja.

"That could be the governor, but I don't think she's involved," said Dylan through a mouthful of jerky.

"How about her senior political advisor? Or anyone high up in the Alaska government. You said a trooper was involved," said Anja.

"We think he was a trooper. He was dressed like one," said Dylan.

"Or it could have been someone posing as a trooper. Someone or a group with lots of money could just hire mercenaries. Who has a lot riding on this scrubber?" asked Anja.

"The whole world does. Specific actors could be someone like Professor White. He has his professional reputation on the line, but he doesn't seem like the type to hire mercenaries. I doubt if he makes enough on his teaching salary to spend a small fortune on hired soldiers."

Anja's voice rose, "What about oil industry executives? There would be a major crash of oil prices if suddenly cheap coal could be burned to produce energy."

Dylan knew she was thinking about her grandfather, "Or it could be an anti-environmentalist like Greg Yeung, or anyone with lots of money." Dylan looked down at his coal dust stained hands.

"Anja, we think there might be problems with *the Scrubber*," said Dylan tentatively.

"What are you talking about?" her voice became still.

"We have unexplained anomalies between the data from the remote perimeter combustion station and what's been transmitted to ARCS. It could mean that someone has deliberately altered the data," said Dylan.

"Or it could be explained logically and nothing whatsoever is wrong with *the Scrubber*," said Anja.

Anja paused as she collected her thoughts. "Let me see if I can summarize the two forces at play here. We'll call the first group, *Pro-Coal* energy. The *Pro-Coal* group is interested in having us believe *the Scrubber* system works; they want coal-based energy plants with a minimum CO_2 footprint. You and your team are working against the *Pro-Coal* group's goals by publicizing *the Scrubber*'s performance problems before the research team can solve them.

"We'll call the other group, *No-Coal*. This group has a strong interest in non-coal energy sources and is trying to stage a public failure of *the Scrubber*. Their thinking is that a media blitz failure will turn both the voters and the government away from coal-based energy.

"You and your team are working against the *No Coal* group's goals by identifying performance problems before the public understands and buys into the illusion of *the Scrubber* system." said Anja.

Dylan shook his head, "So, both groups want us out of the way! What do you propose, throwing in the towel, heading for Mexico and lining up the tequila shots? Hell, for all I know someone completely unassociated with *the Scrubber* project is trying to kill us."

"How are you going to find out if there's been tampering with either *the Scrubber* or the data?" asked Anja.

"That notebook with the preliminary analysis of *the Scrubber* and its specifications is at my mailbox in Seward. We're going to see if we can figure out the source of the conflicting data by digging into *the Scrubber*'s specs."

Sully tossed a nearly empty package of jerky over his shoulder into Dylan's lap. Bits of the dried meat scattered around Dylan just as Gator happened to look around.

"God damn it! You're making a mess in here. You can clean it up when we land. We'll be in Fairbanks in ten minutes."

"Dylan, I want you guys here in my building. I have very competent and dedicated security people who can keep you safe," said Anja.

"Safe? I don't think we have anything to worry about. No one but the miners know where we are heading, and we're pretty sure the killers who were trying to murder us, died in a mine collapse."

THE BOSS ENTERED the complex codes to open the encrypted email system. It was time to send Tango and her team to Seward.

73

DAVE WHITE'S PHONE chirped, and he looked down to see a message from the ARCS proprietary SMS system, *Meet me in my office—Conrad.* Sometimes it irritated Dave that Conrad would use technology when a non-tech communication would do. Conrad would rather send a text when he's sitting right next to you.

Dave walked over to Conrad's office, looked out his office window and noticed the weather had moderated back to seasonally appropriate, breezy and rainy. Conrad seemed uncharacteristically serious as he turned his computer monitor so Dave could see it.

"I was checking the ARCS secure mail site and found an email addressed to you. Well first, you need to know that Ethan's been shot and killed," said Conrad.

"What! How did that happen?" Dave looked shocked.

"Trooper Riggs and some others killed Ethan. I have the location of his body in this message," said Conrad. "That's not all."

"Wait. How could that happen? Did Ethan go crazy? Are they sure Riggs shot him? That makes no sense. He was a trooper for Christ's sake. What else?" yelled Dave.

"Also this, Dylan's away party is not hunkered down in the mine shed waiting out the storm like we thought." Conrad looked grave.

"Well? Where are they?"

"He and the away team are on their way to Seward to review some documentation on *the Scrubber*." Conrad waited for Dave's reaction.

"What? Why the hell would they go to Seward? That makes no sense. There's no data there. We need that coal! How are we going to complete our experiments?" Dave glared at the monitor. "OK Conrad. Wind this back. I want to see everything."

"What I have is a lengthy voice-to-text message from a really bad satellite phone connection. Much of the message is garbled nonsense, but that's about all it says. I think it also says that Dylan will try calling again when he gets a chance," said Conrad. "I'll send the entire transcript to your email."

Conrad thought Dave looked stunned. Dave was apparently processing the murder of Ethan, Zhang's death, the lack of a proper coal sample when Lindsey burst into the office.

"Professor, there you are!" Lindsey stood breathless on the threshold of Conrad's office, a large puddle forming around her boots and snowsuit.

"Hey, you can't bring a wet snowsuit in here!" called out Conrad. "I have sensitive equipment I'm working on."

Lindsey backed out a step and took a deep breath, "Professor, this is urgent. I was just checking on . . . that toboggan we left outside, and I followed some fresh tracks to a huge RV parked about a mile from here."

Dave looked like he was about to explode, "A what? What are you talking about? Damn it! Everything is falling apart!"

Lindsey looked like she was about to cry, "But Professor. That RV. It was just bristling with antennae. I'm wondering if we are being surveilled. Maybe we should get Trooper Riggs to check it out."

"How could an RV get in here? There's barely a road," said Dave.

"It has six wheels. It looks like it's built for off-road travel," said Lindsey.

"Trooper Riggs is unavailable," said Conrad. "I'm calling out for more troopers. Someone needs to check on that RV and this location that Dylan sent us. Maybe the two are connected."

Dave's phone started ringing. He looked down at it and paled, "I got to take this in my office."

Dave stepped out of Conrad's office and walked briskly in the direction of his office.

Harold Cheek's name glowed malevolently on the caller ID screen. Dave accepted the call and heard Cheek's deep voice, "What kind of bullshit is this? Dylan is going to ruin everything. What are you going to do about him?"

74

FROM THE BACK seat of the Cessna 172 Taylor and Dylan watched Sully's head loll back as he fell asleep. He started snoring loudly.

"Who snores with their head back like that?" Taylor asked.

"Maybe a guy who's been shot at, hit with shrapnel, survived a mine rockfall and . . ."

"Your point is taken. We should all snore like that," Taylor said. "What did your daughter say? I could only hear part of your side of the conversation," said Taylor.

"We talked about the evidence that *the Scrubber* may not work as touted. She also speculated on who was behind the attack on us."

Taylor turned to look more directly at Dylan. Her leg pressed against his. "I want to know what she said. It seems like it was the state troopers."

"We discussed that. The governor is the commander of the troopers, but she doesn't seem the type to send them out to kill people. Also, all the troopers I've met were professional law enforcement. They wouldn't take on an assignment to kill a bunch of scientists."

"I agree. It sounds far-fetched to think anyone could just order troopers to kill people. But if it were possible, could the governor do it?" Taylor moved her leg closer to Dylan. He didn't move away.

"Anja and I talked about the governor a while back. She respects her as a shrewd and ruthless politician. Probably anyone who makes it to the top is that way. But she just doesn't seem like a killer."

"How about that guy who's always standing behind her, Harry Cheek?" said Dylan.

"I think his name is Harold. Everyone says he's an asshole, but a brilliant strategist. He may be the Dick Cheney to George Bush." Taylor noticed that Dylan moved his leg away from hers.

"Will someone stop that guy from snoring? I'm trying to operate this aircraft," called out Gator.

"Have you tried ear plugs?" asked Taylor.

"I've tried an entire package, but he keeps spitting them out," said Gator. "There's some duct tape back there somewhere. Just tape him shut. I'm trying to fly."

Taylor turned to Dylan, "What did your daughter say about *the Scrubber*?"

"She seemed a bit upset that there might be problems with it. She wondered if we were safe," Dylan looked at Taylor.

"Are we?"

AT THE ZETROS, Tango answered the secure phone. Despite the voice-altering software, she recognized The Boss and his slow, deep words sounded angry. "I am not pleased with this report. You were only able to confirm one death on the four-person ARCS away team? How's that possible?"

"It should have been easy, but one of the targets you assigned is an expert wilderness guide. He knew how to avoid us." Tango listened to herself and thought she sounded lame.

"How could you let him kill Concord? He was an important asset."

"Concord blew himself up. He did it for the cause. You are sending me fighters who care more about their cause than completing the mission."

"Don't complain about your soldiers. They cost us practically nothing and are willing to throw themselves on their swords for you."

"That's what Concord did. He could have just waited for us to return and kill the targets, but he wanted to get the job done," Tango explained.

"Listen to me. I was told you were the best. Yet you let some scientists and a simple mountain man get away. You blame your failure on some backwoods guy. This can't happen again. I'm transmitting more explicit orders through the secure email system. Once your men are in Seward, I want Dylan and his team dead!"

"Sir, I . . ."

"Don't *sir* me," the deep, mechanical voice boomed. "I don't want to hear excuses."

75

AYUSHMANN KRISHNAMURTHY jumped back under the high-energy assault that was Colonel Bolton. "Oh Mr. Bolton. I'm very sorry to disturb you." He eyed the gun warily as it swung around in a dramatic arc.

"Colonel!" shouted Bolton.

"Excuse me?" Krish took a step back.

"Call me *Colonel* Bolton. What are you doing here?"

"Of course, Mr. Colonel. Can you put that gun down?"

Bolton looked at his revolver as if he didn't know how it got into his hand. He set it down on a nearby side table.

"Did you see the governor's presentation? I have some information you might want to see." Krish peered around the room as if looking for other armed colonels.

"What information could you possibly have that I'd be interested in? Did the governor get hit by a car? I'd like to know about something like that."

"Oh no, Mr. Colonel. My client from Venezuela was very disturbed by the governor's talk. He had his team do some digging on *the Scrubber* she talked about."

"Oh yeah? What did the team find out?" *This little rat has something on the governor.* Bolton felt balance being restored to his world. He would come out on top. He always did. Confidence surged through him. Colonel Bolton was indeed back.

"May I come in?" Krish asked timidly.

Bolton stood back and waved him in. Krish gave the huge revolver a wide berth. Once the door closed, the smell of gunpowder nearly over-whelmed the little man. "May we go into the office lounge, it's so smoky in here."

"Sure. Follow me." Bolton strode out into the hall and downstairs to his office, forcing Krish to trot.

"Alright, Krish. What's so important that you come knocking on my door?" Bolton said.

"Colonel Bolton, I have here something you must read. It's about that scrubber that the governor spoke about."

"Yeah? What does it say?" Bolton looked intrigued.

"It's a report by PTL, a major testing lab here in Alaska, published in an obscure academic journal by a researcher named Dr. Trevor Calm. He has tested a prototype and thinks that the technology can never work on a scale as large as a coal-fired power plant."

Bolton felt a surge of adrenaline flow through him. *God this feels good!* "Good job, Krish. Even if this study is bullshit, we can use it to cast doubt on the governor's claims."

"Yes, we are hoping you will be able to hold up your end of the agreement between my clients and your companies," Krish said.

"There was never any doubt that I would hold up my end. I wonder about some of the slippery devils I'm working with. What do they say?" asked Bolton.

"Is this room secure?" Krish asked, then flinched waiting for Bolton to yell.

"Let me show you something," Bolton said.

He picked up a tablet and moved his fingers over the surface. Krish heard a deep bass thump as all the locks in the office suddenly slammed shut.

"Now look at that monitor," Bolton gestured towards a screen across the room. Krish saw the large screen light up and split into six smaller screens. Five of the screens showed camera views of the office suite. Krish looked impressed.

"I can't even see where the cameras are located even though I think I know where they are," Krish studied the ceiling. Then he stiffened. "Colonel, you didn't record the meeting last night did you?"

"Who do you think you are talking to? Of course I didn't record it," Bolton scowled.

"Can you show me the recordings for yesterday?" Krish asked in a voice that trembled. "I gave my personal assurances to everyone that the meeting would be private. I have some very valuable information to share with you, but I need to know if this room is secure."

Bolton looked annoyed and scrolled through some menus until he came to the time of the meeting Krish had arranged. He opened the file with the meeting's timestamp and saw a still image of himself with all the participants of the recent meeting.

Krish looked up at Bolton with something like contempt. His voice was strong when he spoke, "Play that file." The unexpected order from Krish came like something the Colonel would throw at a private who'd forgotten to salute a superior officer.

"What the fuck?" Bolton looked confused as he hit the play button.

The men watched the video in silence. Krish noticed a sharp intake of breath when it first started, then Bolton exploded. He pointed to the monitor where they could see his tablet on a table next to the Russian oilman. "Look. There's my tablet. I did not start this recording. Someone else did. In this video, I have a drink in my hand, not a control device."

Krish looked doubtful. "Then who started the recording? Who has access to the controls?"

Bolton closed the video file and opened another. "All the key card swipes are recorded in this file." Bolton clicked on an icon of a card reader, and navigated to the night of the meeting. Just a few minutes before the recording started, Jane's card had been swiped.

"You have a traitor in your organization," Krish said. "What are you going to do about it?" Krish let his eyes wander over to the Colt.

"I know just what to do." Bolton's eyes flamed as he spoke.

76

IN THE SOFT LIGHTS of the Netsvetov Hotel bar, Jane ignored a sparkling cranberry-colored drink and listened to Shannon's soft Texas accent.

"Now Jane. Are you sure Bolton was lying to those oil guys when he said he broke the law?" Shannon sipped her water-and-lime holding the straw with delicate fingers.

"The thing is . . . I don't know for sure. Hell, I really want to believe him! Bolton is smart. If he's lying to me about his criminal behavior, he's setting me up. It's possible he is really pissed at me. It was my responsibility to implement our agreed-to strategy, but within those limits I made several really bad decisions.

It's my fault the Super PAC has been moving in the wrong direction. I should have limited Bolton's ability to freely spend money. This could ruin him."

"Is he talking about getting revenge on you?" Shannon asked.

"I haven't talked to him since that night," Jane said. "But when things go wrong, he scatters blame like a leaf blower loose in a confetti factory."

"But it's not your fault. Who would have figured that the governor's team would have this coal and carbon dioxide thing ready to pull out of their butts? What's Bolton going to do? Fire you?"

"He might do much worse. He could ruin my future career. I could end up back in Nacogdoches gutting chickens in the slaughterhouse." Jane stirred her drink.

"I need to help him in order to save my career. He told me to go back and get more information from my source in the governor's campaign. But I don't think I could talk to Cheek again."

"Don't take this lying down. Let's find out if that scrubber thing actually works. Maybe it was made up by the governor and sprung so late in the election as a trick."

"That doesn't seem consistent with what I know about the governor," Jane said. But the idea of doing something made her feel better. "If that scrubber is what the governor says it is, then we are screwed. If it's a fake, we'll need to mount a very vigorous campaign to get the word out. The Colonel has put his all into this campaign and really needs the win. If he wins, so do I. I'll go on to an even bigger campaign."

Shannon stood, "Grab your drink. Let's go into my office and do a little research. Even if we can cast some doubt on *the Scrubber*, you might be able to salvage your job," Shannon paused. "Helping Bolton makes me feel like a political mercenary."

"Tell me about it. I'm feeling that way more each day," said Jane as she stood next to Shannon.

Hours later, both women were clicking away on keyboards in Shannon's modest private office.

Shannon glanced up at a wall clock. "This isn't working, and I need to make an overseas call in 20 minutes."

"OK. Let's look for another few minutes, then I'll leave so you can make your call."

"Why don't we try looking up the names of the people involved with *the Scrubber* to see if we can find out something about them. Maybe one of them is a relative of the governor," said Shannon.

"Here's something," Jane pointed to her screen. "An Alaskan guy, named Dylan Baker, first patented the process to remove CO_2 using bacteria-laden dirt. That was over 30 years ago though."

Shannon gazed at her monitor and clicked her mouse, "And here is a more recent article by a researcher named Dave White."

"Whoa! I looked up that Dave guy, and this article says he had Partikkel Tek Labs test a prototype unit. PTL is right here in Anchorage!" said Shannon.

"That's a connection alright. Let's see if we can find a relationship between Partikkel Tek Labs and that bacteria *the Scrubber* uses," Jane said.

After another few minutes of searching, Shannon squealed, "Look at this! Here's a report about *the Scrubber* from Partikkel Tek Labs. It's behind a pay wall. We have to be able to get into this expensive academic research database in order to see the report. It costs $3,400 for a one-year subscription. No wonder no one knows about this report."

Jane pulled her personal credit card out of her wallet, "Let's see what it says."

Thirty-four hundred dollars and minutes later, the two women read the report by Dr. Trevor Calm and gasped at a line in the summary: *the Scrubber performance is*

*being impacted by intermittent fluctuations in CO_2
removal. More performance data is needed to properly
assess industrial potential, but as it stands, the
Scrubber is not an effective CO_2 mitigation device.*

"I know how to save my reputation and help the
Colonel! Maybe my career isn't over yet," said Jane.
"But it depends on if I can get this Dr. Calm guy to talk
to me."

77

JANE KOSS PAID for her coffee and dodged the
tables and patrons to a small counter to add milk. She
loved the way Alaskans revered exceptionally good
coffee at all hours of the day. She doubted she could
ever go back to the bitter, nasty stuff she drank in
college.

Her phone buzzed; it was Bolton. He would undoubtedly chew her out in his military voice just before firing
her.

"Hello, Sir," her voice was tentative.

"Jane, it's me."

She would have expected an irate tone and instead she
heard a surprising warmth and peacefulness in his
voice.

"It's been a crazy evening! I'm still trying to sort it all
out. I have a lot of things I want to cover, so give me a
moment to run through my list."

Jane's mind was spinning from Bolton's calm tone
and engaging manner. She had expected all hell to
break loose. "OK."

"I'm sure you heard the governor's speech. I'm as
upset as you must be. Yes, I know that focusing
Yeung's campaign on stopping coal development was
your idea. But don't beat yourself up over it. There was
no way we could have anticipated a new device that
decreases CO_2 emissions or some freakish pure coal

seam? Get past those feelings. I have an idea that may get us back on track."

What's with Bolton? This whole speech was exactly the opposite of what she had been dreading. Both relief and anxiety washed over Jane.

"Last night I had a meeting with a bunch of big oil guys at the office."

"Yes, you mentioned the donor meeting without including me." *Why was Bolton admitting this?* Jane thought. "I heard about the armed guards in the stairwell."

"I know; these guys can be a little melodramatic. They had requested that I keep this meeting confidential."

"How did the meeting go?" asked Jane, not knowing what to say.

"Great. But, I had to lie to get them to agree to work with me," said Bolton.

"You lied?" asked Jane.

"Yes. I . . . get this, I told them I borrowed money from the Super Pac and committed stock fraud in order to get enough money to make a deal with them. They totally fell for it."

"But you didn't actually do those things?" asked Jane sounding more upbeat.

"Of course not! I am not a criminal. I told those guys what they wanted to hear so they would cough up some money for the Super Pac," said Bolton. "I need you to keep this confidential. If the governor's team heard about the meeting, they would blow it up all out of proportion."

For a few moments, Jane's mind simultaneously considered how close she had come to destroying everything. She had been so close to tossing in the Bolton towel and accepting Cheek's offer. Now she felt guilty. Loyalty had always been one of Jane's key life-principles.

Bolton continued, "I need you to help gather some additional information. I think we can pull this mess out of the fire."

As Jane listened, she felt as if the old, brilliant Bolton tactician was back. Whatever had blocked access to the Colonel's strategic skills had been shoved aside. Rather than being pissed, Bolton was giving her a chance to save her job and get a great recommendation. Jane felt a tingle of excitement. Maybe her career as a political campaign expert was not dead.

As soon as Bolton finished talking, she snapped a lid onto her coffee and emerged from the coffee shop knowing whom she needed to see next.

COLONEL WILLIAM BOLTON leaned back in his office chair and gazed out into the Anchorage skyline. *That went very well!*

78

ANJA LOOKED OUT of her floor to ceiling corner-office windows at the Chugach Mountains rising up from deep forest greens to dramatic snowy peaks. Looking the other way, she gazed at Cook Inlet for a while and watched the ever-changing clouds and water traffic, including several floatplane landings. Alaska was her new home, and she already felt a powerful loyalty to it. She didn't easily offer her emotions to anyone or anything.

Two years ago, while working for GreenWorld, she found a tiny part of herself that allowed her to fall in love. Milo packed a lot of man into a smaller frame. Milo lived life with such fierce joy, that Anja viewed him as a force of nature.

They met on *The Sylva*, a 321-foot former Canadian icebreaker retrofitted to allow direct action environmental protests. At that time, it was tasked with disrupting Arctic oil drilling.

Milo was an electrical engineer working onboard, and Anja was escorting big celebrity donors on an Arctic mission. After that first meeting, Anja found herself spending months at a time on *The Sylva*. She tentatively allowed herself to fall in love. It was a huge risk for her, because she had spent most of her life walling herself off from others.

The euphoria of loving someone completely overwhelmed her inhibitions. She found herself softening all of her relationships and becoming interested in socializing with her staff and their families.

Then an oilrig called *Arctic Sun* ran aground on Andronica Island in the Gulf of Alaska while holding nearly 150,000 gallons of petroleum products. It was the kind of rig that was towed out to a drilling site to pump oil in the summer and hauled back to civilization during the harsh Arctic winters.

The Sylva's mission was to disable the *Arctic Sun* so it could not be used for any more drilling, but to avoid spilling the oil in the process. The plan was to damage the double-sided funnel-shaped hull, so it would no longer be of any use in icy waters. The company would have to tow it to China to be scrapped. This could cost the oil company millions and delay their drilling plans in the sensitive Alaska waters.

Milo and a demolition team boarded the oilrig to place explosives when suddenly the rig listed, and a loose 500-pound drilling collar crushed Milo as it rolled across the deck.

To the people around her, Milo's death didn't appear to affect Anja in the least. However those closest to her noticed she became quiet and emotionally distant. No one saw her cry or mourn. Inside, she realized her survival depended on her ability to bury her feelings deep under a crust of cold logic and passion for her work.

ANJA HART looked down at the caller ID on her office phone. It was the governor's senior political advisor who, rumor had it, was attempting to arrange a substantial campaign contribution from the owners of the *Arctic Sun*. She did not want to take this call.

"Ms. Hart, it's Harold Cheek on line one. Should I take a message?" Roger had entered her office with a tray of vegetables and hummus dip.

"No. I can take the call," said Anja. She picked up the handset and pushed a button, "Anja Hart here."

"This is Harold Cheek. What's with the hesitation of the Mountain Club to endorse the governor? We had this all figured out, but nothing's happening."

"We've gone over this. We are waiting for the governor to openly state she opposes offshore and on-shore oil exploration in new areas. I could ask you, *what's the delay?*"

"Look, you need us. We are your best shot at pre-venting those new oilrigs appearing all over the coast like quills on a porcupine. If that asshole Yeung gets into office, you'll look very ineffective with the coast bristling with oil rigs."

"I need you? If that's the case, why are you calling me? I'll tell you why. If the Mountain Club endorses Bates-Hardy, we can offer millions in campaign money and hundreds of volunteer campaign workers. You need us far more than we need you." Anja hated Cheek and his bullying.

Cheek changed his tact. "By the way, what's all this bullshit that your dad is spreading around? He called our office to tell us *the Scrubber* might not work and that someone is shooting at ARCS scientists. This is total unsubstantiated crap. Can't you control you own backwoods dad? Give him some moose-jerky, so he'll shut the fuck up."

"He's not my dad. I just met him recently for the first time. He's a sperm donor. That's all. Don't put his misbehavior on me." Anja dipped a carrot stick into some garlicky hummus. She held the mouthpiece of the phone away from her as she crunched.

"Well we have word that he respects you. As a dumb hick, he might just do as you ask," Cheek urged. "We don't need rumors about *the Scrubber* not working getting around."

"He's no dumb hick. The Scrubber is based off his original designs, and he understands it. He called me with the same story about some bad guys, including one of your troopers, killing or attempting to kill ARCS team members." Anja found herself pointing a piece of celery at the phone as she talked.

"You put him up to this, didn't you?" Cheek said. "If *the Scrubber* doesn't work, it strengthens your bargaining position. You are playing dirty." Cheek's voice rose.

"You probably put him up to it, you asshole! How dare you accuse me of this!" Anja threw the celery across the room.

"We need to meet," Cheek said. He spoke like he expected everyone to just drop what he or she was doing and run to him. "And don't try to get out of it. We need to hash out your endorsement and a timeline. I want your word that you're not going to encourage your idiot dad to screw this up for us."

"Why the hell should I meet with you? You need me more than I need you."

"Because I'm bringing the governor's complete drilling moratorium proposal for you to get a secret peek at. I don't know why, but she wants your input."

Complete moratorium? Whoa, that's new. Anja thought. "OK, when and where?"

"Meet me at the Botanical Gardens on Friday at 4 pm sharp. Come alone. If you bring that Roger goon or anyone else, it's off. We're pulling back the moratorium." Cheek hung up leaving Anja alone in her office.

"Bitch!" said Cheek to his empty office. She won't be pulling the strings for much longer. He punched in a number on his phone and prepared to give orders.

79

BOLTON PLACED his antique Colt revolver back in its case, his mind not even acknowledging that he had nearly taken his own life with it. A man unburdened by self-reflection, Bolton looked ahead to a bright future with confidence.

Anyone observing might think Colonel Bolton was able to command loyalty from his troops by sheer personality. He conveyed mountains of personal magnetism that made nearly everyone he met awestruck or eager to please. But that wasn't all he had, he instinctively understood people and their motivations.

Bolton thought about his call with Jane. He had played Jane like a violin. By the end of their phone call, he had manipulated Jane so she was certain that he had not broken any laws, and her best interests were to continue to help him.

She probably believed he was her ticket to lifetime success so she couldn't let him down. Maybe she could get some information about *the Scrubber* from the governor's campaign that could cast doubt on its effectiveness. If anyone could, she could.

His tablet made an alert sound. He picked it up and smiled to learn that Jane had entered her office. She was back to work for him.

AS JANE BOOTED up her computer, she looked ahead to what she would do when this campaign was over. After getting Yeung into office, she would be rewarded with access to Bolton's extensive network of major political connections.

Somehow she felt dirty, as if she had to put aside her personal beliefs in order to pursue her career. She remembered what the manager of her first campaign had said, *Lawyers defend guilty people because every client deserves their day in court. As professional campaign workers, we might serve a weak candidate because every voter deserves to have alternatives at the ballot box.*

Opening her browser, Jane did a search for the author of the Partikkel Tek Labs report. What came up was *Dr. Trevor Calm.* He was more than a research analyst at the famous, directaction environmental organization. Calm was the senior director of research for Green-World.

He supervised Partikkel Tek Labs. PTL did independent testing for academia and companies as well as gather information for lawsuits against anyone willing to damage the earth or challenge GreenWorld. It also appeared that GreenWorld attempted to disguise their relationship with PTL. Maybe this was to remove any appearance of bias; maybe it was just the well-known institutional paranoia that plagued GreenWorld.

Even saying that the PTL did some of the testing for *the Scrubber* could cast doubt on its performance claims. Jane felt she was really getting somewhere. If she could discredit *the Scrubber*, Yeung could still get elected. She wanted to talk to someone in the lab about the report, Trevor Calm if possible.

The more she dug into Calm's background and GreenWorld's tactics, the more she became confident that some of the testing on *the Scrubber* could have been manipulated or hidden. GreenWorld's motivation could have been just to draw attention to global climate change. It seemed they would do anything for attention to their causes.

How would she ever get to talk to Trevor Calm if he worked for one of the most paranoid activist organizations in the whole state? GreenWorld was like an insular religious organization with secret meetings and armed guards around their higher ups.

Lots of people in the energy business thought only crazies worked there. However, as she did more searches about Calm, she noticed he had gotten his undergraduate biology degree at University of California, Santa Cruz and his first doctorate in environmental policy at Bard College in New York. The guy was smart and liberal.

Maybe she could just make an appointment with him. She picked up the phone and called the number for PTL. After navigating a cumbersome phone directory, Jane got a human.

"Partikkel Tek Labs, may I help you?" came the answer.

"Yes, this is Jane Koss at University of Alaska, Anchorage, may I speak with Dr. Trevor Calm?"

"One moment, please," came the reply. A few minutes later the operator came back, "May I ask what this is concerning?"

"Why yes, we were replicating some of the tests PTL did on a carbon dioxide metabolism device called *the Scrubber* and came up with disparate results. I'd like to talk with Dr. Calm about testing protocols, but it would be best if we could meet."

"One moment please," repeated the operator. It was quite a few minutes before she came back, "Dr. Calm asks that you send over the protocols used by your lab as well as the results so he can review them. He's sure the answer is a simple break in procedure."

"I'll bring them right over," said Jane. The operator gave Jane an address different from the one on PTL's website and access code to get into the parking lot because some of the security staff would be mostly gone by the time she got there.

Jane placed her voice recorder into her pocket, scooped up a sheaf of blank papers and shuffled them into a manila envelope and went out to her car. She felt uneasy about the subterfuge and hoped GreenWorld's reputation for swift retribution was unearned.

80

"EVERYONE, BE QUIET until the prop stops. We're coming into the Fairbanks airport, and I don't want any distractions," Gator barked.

Sully sat up and started fiddling with his data recorder. He looked around, "That was a fast trip. How far did we come?"

"What part of 'be quiet' didn't you understand?" Gator glared at him.

"I was just wondering how far we came." Sully looked down at his recorder. He entered a value and it beeped. "I thought so!"

He turned around to face Taylor. "A couple of weeks ago you and I were discussing *the Scrubber*'s injection port configuration. What was your comment?"

Taylor paused thinking back to the conversation. "Yes, I remember. I asked you whether you recognized the equipment just outside of the lab. It was a cylindrical container just upstream from the CO_2 sensors and *the Scrubber*'s injection port. I hadn't seen it before and asked if you knew anything about it. You gave me one of your famous 'blank looks'."

"True, I did. But now I may have an explanation," Sully beamed slightly.

"OK?"

"Shut up you guys," growled Gator.

"Well, the ARCS data we've seen up to this point indicated CO_2 levels at *the Scrubber*'s injection ports are many time higher than post-*Scrubber* measurements. High concentrations going in and low coming out, and everyone was happy!" Sully pointed to the recorder screen. "But this measurement says that the CO_2 levels at *the Scrubber*'s injection ports and *the Scrubber*'s exit ports are the same!"

"What the hell?" Taylor stared out the window as she processed this new information. "Are you telling me that the CO_2 levels from the combustion station were lowered *before* the gas entered *the Scrubber*?"

"God damn it. Shut up!" Gator exploded.

"Correcto-mundo! In 2009 some Canadian company proposed using a solution of sodium hydroxide to capture carbon dioxide. I suspect that if we opened up our little cylindrical container, we would find it brimming with sodium hydroxide. All the bastard would need to do is to electronically replace *the Scrubber* injection data signal with a pre-sodium hydroxide treatment signal. We would not know that the exhaust gas had already been scrubbed of CO_2 before even reaching the Scrubber system! The decrease in CO_2 levels we observed was unrelated to technology. Pretty slick, huh?"

Dejected, Taylor and Dylan both stared out their respective windows. Sully sat back into his seat feeling both mentally successful and ecologically depressed.

"I'm going to break up someone's testicles if he doesn't shut up," Gator threw a middle-fingered gesture at Sully.

"Cessna 5-8 juliet approaching 2.5 requesting 3 thousand," Gator spoke into her mic to the control tower as she glared at Sully.

"When I first started working with that bacteria, I did not understand that it was alcohol that was killing my germs," said Dylan softly. "Maybe Dave's alcohol removal system doesn't work as well as he thought."

Sully turned in his seat, looked at Dylan and stated in his normally loud voice, "Are you saying Dave doesn't know what he's doing, or that he is deliberately sabotaging his own work?"

Just then the radio came on, "Tower to 58 Juliet. Go to three thousand."

Gator thumbed the mic, "Shut the fuck up." Then she looked at Sully, "5-8 juliet, Roger on that."

"Tower to 5-8 juliet, repeat please."

Gator turned pale. "Listen you guys. You really need to let me concentrate, OK? I just told the tower to shut the fuck up. They could pull my license for that." She thumbed the mic, "5-8 juliet, Roger on that."

At once, the passengers seemed to realize they should be quiet while Gator handled all the tower communications and landing procedures. They remained utterly silent until the plane had landed and taxied up to the general aviation part of the airport. Soon the prop stopped.

Gator seemed visibly deflated. Her miscommunication with the tower changed her typically robust verbal style to something quiet and tentative.

"Guys, we'll be here an hour. I need to pee, refuel, send off these ore samples, and call my daughter. You can walk over to the main airport and get a meal or a drink," with shoulders slumped, Gator opened her door and left the plane.

"I think we screwed up," said Dylan. "Her confidence is shaken."

"Let's buy her some candy bars," said Sully as he exited the plane.

"No," said Taylor. "We need to apologize and make it up to her."

"How do we do that?" asked Dylan.

"I don't know. Let's grab some real food and figure it out." Taylor pushed the seat forward and exited the small plane. "I want to look at those recorder data so we can find some agreement."

"And I want to check my email for something from Suzie. Taylor, can I use your phone?" asked Dylan. He found a table by himself and punched a love email and sent it off. His email inbox was woefully empty of anything from Suzie.

Sitting at the airport café over hamburgers, the team had decided to buy some airport baby clothes for Gator's expected child, and they had each signed a card that Sully picked out saying they were sorry they had distracted her.

Sully stole a french fry from Taylor's plate and said, "So I've been thinking about what you and Dylan said. And I went over these data again."

"And you've decided Dylan and I are right," said Taylor confidently.

"I'm not sure what you mean by that, but I do agree that it's very likely *the Scrubber* had some kind of extra power to do its job." Sully squeezed the nearly empty ketchup over his opened hamburger. It made a loud farting sound.

"Say *excuse me*," teased Dylan. "If we all agree that *the Scrubber* was given more power than reported to the ARCS computers, then it's possible that it's a fraud."

"I hate to say it, but it's pointing that direction," agreed Sully. "In the end, it doesn't really matter how *the Scrubber* disassociates carbon-dioxide if it's using too much power to have practical application."

A big hand slapped Sully on the back, "OK you eggheads, time to board the plane. Same rules as before, no throwing up. New rule, don't talk when we're preparing for landing." Gator seemed back to normal after talking to her daughter.

81

JANE DROVE HER SUZUKI X90 up the gravel drive to the GreenWorld Headquarters. The dark and cheerless stone building stood under a dreary overcast sky. With decorations of battlements, stone spires and a slate roof, the headquarters building was constructed mostly in 1922 as an insane asylum.

Until this building was erected, Alaska arrested its citizens for the crime of being mentally ill and shipped them to Portland, Oregon for incarceration. In the 1960s, the budget-busting building on the outskirts of Anchorage was decommissioned and allowed to further decay.

The 1970s testing of nuclear weapons in Alaska energized dozens of environmental protest organizations, but the one with the most over-the-top direct actions was GreenWorld. The organization was as much about publicizing their protests as anything else. They skillfully chose charismatic, symbolic animals, like whales and dolphins to put into their press releases and funding requests, while spending as many resources saving the lowly krill.

As a result of their effective publicity activities, donations poured in. GW bought the former Alaska Asylum for the Insane and made it into their headquarters. Locals sometimes joked about the crazies at GW Headquarters. As many of their supporters matured, donations of money and time shifted to the less dramatic more socially appropriate Mountain Club.

While many in the environmental movement disliked the gritty, in-your-face tactics of GW, it was obvious that the combination of overt, even outrageous, protests along with publicity was getting the attention of voters and lawmakers.

Jane parked her X90 in the nearly empty visitors parking area. She noticed a security camera on a pole tracked her movements as she walked past blast walls and approached the large doors in front of the building. A large, heavy oak door, that looked like it had been stolen from a European castle, swung open as she neared the entrance.

Another door, very modern with thick glass, chrome and decorated in gold with the GW logo, slid open to allow her into the security-monitored lobby. Once there, the vibe of the building changed from 1920's dreary fortress, to environmental chic. Classy recycled furniture and vibrant photographs of *The Sylva*, marine mammals with sad faces and action-shots of protests decorated the walls.

Jane presented herself at a reception desk and was greeted by a large woman who looked as if she were a hippy 40 years ago with a tent-sized tie-dye mumu, gaudy metal jewelry and large black-framed glasses.

"I'm Jane Koss. Dr. Calm is expecting me." Jane smiled up at the woman who had an expression as if her feet hurt.

The woman picked up a tablet and scowled at it. "Hi, my name is Evelyn. Yes. He put you on his list. Have you been here before?" Evelyn asked.

"Nope. This is my first time."

"Well, you need to get a bracelet," the woman held up her wrist to show a slender metal high-tech-looking smartwatch on her wrist. The receptionist reached under her desk and pushed a button. Across the room a door popped open. "That's security. They'll check you in and give you a bracelet."

Jane had an uneasy feeling walking through the door that had just opened; it had no doorknobs and a metal plate on the inside. She felt as if she were entering a prison cell when the door clicked solidly shut behind her. Once inside, the prison feeling was greatly magnified by a mixture of thick glass, actual metal bars and a small manufacturing lab. Here there was no relaxed hippie feel; it was all grim efficiency.

A young, bald man with wire-rimmed glasses and a white lab coat came out of a side door and introduced himself as Jube, a security tech. He wore the same metal wristband that the receptionist showed Jane. For the next few minutes, Jane had her retina scanned, driver's license photographed, fingerprints taken and she filled out a security questionnaire on a tablet device.

Jube looked at his tablet and said, "You made a mistake. You said you worked at University of Alaska, but it says here you work at Prosperity for Alaskans."

Jane blushed. "That's wrong. Please change it."

Jube's fingers clicked on a keyboard, "I flagged that fact for further research. You also said you own your car, but we see you still have six payments left."

Jane's embarrassment changed to anger. "You don't need to know all that so I can chat with Dr. Calm."

"Well, you checked the box that said you agreed to our terms and conditions," Jube looked at her seriously. "Would you like to leave now?"

Jane thought of her career taking off after she completed her current job. Really, her future depended on her ability to please Bolton. "I'm fine," she said.

The check-in was so thorough that Jane wondered if they'd require a body cavity search before letting her past security. Finally Jube opened a locker where she could put her purse and jacket, then showed her a metal bracelet. He locked it tightly onto her wrist with a tool that made her think of a police handcuff.

"We'll take this off before you leave the building. You actually can't physically leave the building with the bracelet." Jane wondered what would happen to someone's wrist if she jumped out a window.

"The bracelet will have a green pulsing light if you are in an area you are cleared for. It will pulse orange and vibrate if you take a wrong turn. If it turns red, you have entered an unauthorized area. Stop and push this button on top. Security will come.

"If you don't stop, the area you are in will go into lock down until security can find you. I don't think you need to worry about the red light. Your bracelet will not open any unauthorized doors."

"Fine. Where do I find Dr. Calm?"

"Look at your wrist," said Jube.

Jane looked down to see a text message on the small screen: *Exit blue door.*

82

AS SHE STEPPED OUTSIDE of the Anchorage Governor's Mansion, Governor Leola Bates-Hardy took a deep breath of the fresh Alaskan air. She realized she was bit of an anomaly, an Alaskan who never tired of the crisp, clear spring nights. It was 4 p.m. and the evening sky had already settled in.

Even outside Leola could hear occasional excerpts from the celebrations from her campaign staff. The media's response to her presentation had exceeded everyone's expectations. National newspaper headlines captured attention with "Humanity Evolves" and "Alaska to the Rescue". Each article summarized how a new technology would significantly reduce carbon dioxide levels in exhaust emissions, thereby reframing human civilization.

Leola had truly enjoyed seeing the replay of the presentation video. The re-election team had released the Oslo Arctic Climate Conference video several times to many news outlets, websites and blogs. She watched a video stream on her phone for the second time.

"The Alaska Remote Climate Station has successfully evaluated a prototype design that dramatically reduced carbon emissions from coal-burning power plants."

Bates-Hardy paused and smiled. "And, this technology has environmental impact implications for any CO_2 emitting industrial processes."

Leola's thoughts jumped back to the debate at the University of Alaska. She could still picture the re-election team's ear-to-ear grins when Bolton had said he had no doubts that any kind of effective carbon dioxide mitigation device was decades away.

Leola stepped back into the mansion immediately reconnecting with the celebration. Leola looked through the conference rooms for Cheek. It was very unusual for him to be absent from the pulse of the team. Given Cheek's contribution, she was surprised not to see him sharing in the celebration.

The news report on the television was summarizing the implications of the governor's announcement on the upcoming election. Election politics sometimes seemed to change as quickly as the weather near Denali. Within hours a chilly spring storm would blow in making all Alaskans shiver.

83

PROFESSOR DAVE WHITE set *the Scrubber*'s performance data on the desk and stretched. *I need a cup of green tea.* Dave realized how good it felt to move around. And besides, his eyes needed a break from the hours of papers and computer screens.

He reviewed his options. Maybe the Gunpowder tea today; the other green teas, like Sencha, were too light. He needed a little more caffeine, especially after that call from Harold Cheek.

"Anja is a loose cannon; we don't know what she'll do next," Cheek had said. Dave White knew that was an odd assertion. Mountain Club and carbon emission scrubbing technology should be aligned toward the same end, a better global environment.

It was true that more coal burning plants would mean more sulfur and other particulates, perhaps less of an issue with the super pure coal seams that lack heavy metals. Dave shook his head as he compared the two sides of the argument. He decided that one needs to compare the coal-based outcome with the environmental impact of oil-based energy production. Dave wanted to stretch.

Passing a hallway window, Dave glanced at the snow swirling around the ARCS landscape. The storm had dramatically increased over the last couple of hours. Its noise and intensity were already a distraction for the staff. The morning news had predicted that this storm would encompass everything between Anchorage and the ARCS facility by this evening.

Cheek was worried about Anja Hart. She had been tracking Dave White's research for years. Lately she had been hounding White, trying to get him to elevate the priority of specific environmental projects. She was always challenging White and often questioning the ARCS's technical conclusions. Hart had characterized the ARCS facility's first few months as an "introductory lesson in disappointment and disillusionment".

Yes, the ARCS facility started out on a dubious path. *Yes*, Dave thought to himself, *some of the initial ARCS research results could not be duplicated by other labs*. Dave had worked hard to change the policies and procedures responsible for that lapse; both the ARCS facility and his team were beyond that now.

Unfortunately, Dave now found himself being pushed by his twin goals to improve ARCS' reputation and validate *the Scrubber*'s performance. The edge of a GreenWorld envelope peeked out of a stack of papers. *What the hell?* Dave quickly shoved the envelope back into the stack of papers and then checked the hall to see if anyone noticed his odd behavior. *I can't open that letter now!*

To accomplish his goals, Dave might need to temporarily stray down a path of deception. The money in that letter must stay hidden. He would use the Green-World money for the new ARCS equipment they desperately needed. The new equipment would re-energize the team and quickly move the project toward his goals.

Dave refocused his mind on *the Scrubber* data. He began recalculating the recent performance averages after selectively excluding some of the unacceptably low results. He needed to exhibit his steadfast belief in the technology. There could be no distractions.

Back in his office, Dave wondered, *What would Anja do with the real Scrubber performance data? Would she place long-term environmental goals over short-term political gains?*

The golden cloud flowing from the Chinese Gunpowder tea bag as it repeatedly plunged in and out of the water in the teacup seemed to represent Dave's determination to keep addressing *the Scrubber*'s problems until they were solved.

After seeing the University of Alaska conference, Dave tried to figure out why Bolton was so anti-coal. Dave's smile morphed into a grimace as he pondered Bolton, a ruthless enemy with the power and political resources to change the course of nations. Bolton's disgust with both the coal industry and any environmental program was legendary.

Grateful for Gunpowder's elevated caffeine levels, Dave sipped his tea and thought about his retirement. It was all invested in coal. He'd be rich.

84

AS THE LITTLE GREEN and white Cessna 172 labored to lift off from the Fairbanks airport, all the occupants, save Gator, started to nod off. Dylan, sitting behind Sully, had so little legroom that he would not be able to obtain deep sleep. It wasn't just the cramped cabin; a nagging doubt would prevent Dylan from slumber. After a while he started examining Sully's data recorder.

"Sully, are you awake?" Dylan had Sully's recorder in his lap. He spoke to be heard above the engine and wind noises.

"He's asleep. Don't wake him up!" Gator gave Dylan a withering look.

"I'm awake," said Taylor.

"Something about Dr. White's redesign has been bothering me," Dylan said. "I get it that *the Scrubber* seems to initially work just fine. And I get it that when we get to Seward and examine the preliminary test and design information, perhaps we can explain the high electrical power usage."

"OK . . ." Taylor encouraged Dylan to go on.

"I understand that he uses bacterial enzymes in a multi-step process to change CO_2 into alcohol."

"But . . ." Taylor prompted.

"But what part of a multi-step, bacterial enzyme reaction requires significant amounts of electrical energy?"

Taylor took the data recorder and studied it. "Damn. That is really weird. Sully's been thinking about it. Maybe he's had some insight."

"Don't wake up the beast," Gator said. "He gets on my nerves."

"He picked out the baby gift," Taylor said.

"Yeah. It's a nice UAF baby sweatshirt, but I hope my kid doesn't go to the University of Alaska, Fairbanks. She'll be a Florida Gator like me."

Dylan realized Sully's mistake. There could be no doubt about Gator's college loyalty. He should have picked up the *I left my heart in Fairbanks* shirt.

Sully stirred. "I should have bought the *I left my heart in Fairbanks* outfit, but it cost a dollar more."

"Now that you're awake, any new thinking as to what process within *the Scrubber* might require significant electrical energy?" Taylor tried to suppress a grin.

Sully was using his index fingers and thumbs to hold his eyelids open. "You are addressing the early-in-the-process energy spike, right Taylor?"

Seeing Taylor nod he continued, "There are energy processes, like the use of a powerful laser, that can tear apart CO_2 molecules. But we already know that *the Scrubber* process is flawed, so for the moment let's assume that the energy spike somehow enhanced *the Scrubber's* performance."

"Interesting," Dylan speculated. The use of a laser could explain the initially low CO_2 levels in the exit gas even though the bacteria were not growing. And, that would also explain the power consumption."

"These conditions would make *the Scrubber* worthless because of the steep energy requirements to bust up the CO_2," said Taylor.

"OK. Everyone shut up. We're coming into Anchorage air space," Gator said.

Sully turned in his seat to face Taylor. "I hope Dr. White has an explanation for this. Otherwise, with the extra energy consumption, all of our work on *the Scrubber* is a waste of time."

"Are you going to do this again? Are you going to distract me?" Gator growled. "You can shove your UAF sweatshirt if I hear another peep."

Taylor silently handed Sully a Snickers and indicated he should put it in his mouth.

85

AS JANE APPROACHED the blue door, it swung open revealing a dark, but grand marble hallway. The bracelet pulsed blue, vibrated and displayed a message: *Turn left to elevators.* Jane saw a drinking fountain on her right and took several steps toward it. The bracelet vibrated again, this time pulsing orange: *Wrong way. Turn left.*

She couldn't ever remember being so intimidated by technology. Deciding that drinking fountain would not work unless it had been programmed into her bracelet, Jane decided just to follow the directions on the small screen on her wrist.

Standing in front of the elevators, Jane noticed there was no call button. However after a moment, she heard an elevator bell and the doors slid open. When she entered the car, it surprised her that there was no way to tell the car which floor to go to. She looked down at her wrist to find out what was expected next, but no message appeared, just a pulsing blue light, the doors slid closed and the car lurched.

Instead of going up, she felt the car descending rapidly, almost like a mad carnival ride. The car quickly slowed and allowed her out. Here the hallways had no marble. They appeared more like a ship with linoleum floors, metal walls and a ceiling striped with pipes, conduits and wires. To her left, the hallway appeared to disappear into infinity as it led laser-straight into blackness. On the right, overhead lights flickered then came on too bright. She looked at the bracelet: *Turn right.*

The air felt cool but heavy, and she couldn't stop a feeling of being trapped in a deep underground prison. No one knew she was here, and she had no way to communicate with the outside world. They could just seal her up in a room and no one would know.

As she walked, the lights behind her winked out. Her shoes made a lonely echo that could have been used in a horror movie. She passed door after door as she proceeded, none of them with doorknobs or labels. When a door next to her opened abruptly, she nearly wet her pants. A small, gray-haired Asian woman wearing a lab coat and carrying a tablet seemed surprised to see someone else in the hallway. The woman said nothing but turned to go the way Jane had come.

Jane looked past the woman to look into the room she had just exited. It looked like a metal high school gymnasium with a golden hardwood basketball floor, bleachers and lockers. As the woman wordlessly left the room and turned toward the elevators, the gym room lights shut off and the door closed with an authoritative click. Jane sneaked a look at the retreating woman, and noticed her wrist glowing with the pulsing blue light.

This place had Jane so off-balance, that she wondered what she would say to Dr. Calm. She had rehearsed a story that seemed plausible, but dare she lie in this place? The bracelet was probably a lie detector that monitored heart rate and skin reactions.

She felt a buzz at her wrist and checked the bracelet: *Next door on your right.* When the door clicked open, Jane composed herself and stepped in.

Trevor Calm was just turning around when she walked in. He was dressed in a white lab coat, dark slacks and polished wingtips. Jane got the impression that he wore a suit jacket when not wearing the lab coat. To her, he looked like Rand Paul, but much darker. She wondered if one of his parents had been black.

"Jane Koss. You are a slow walker. My bracelet said you were expected here a couple of minutes ago."

"I stopped at the gym for a workout," she said.

He looked confused. "There's no gym on campus. And you didn't need to hand-carry the documents. Next time just email them or upload them onto our website using our file transfer protocol software."

"Why isn't there any PTL logo on the front of this building? You guys are a major lab," said Jane.

"We do some basic lab work in our front office downtown. It's our official address, but here we have our most advanced equipment. The world doesn't need to know that PTL is owned by GreenWorld."

Jane held up her security bracelet, "Nice. I always wanted to get one of these smart watches so people would know I was running late," Jane said as she looked around the lab.

The lab was not what she expected. In the large classroom-sized lab, there were six workstations with lab stools, sinks, small refrigerators, power outlets, gas jets and not much else. In an adjoining room she could see pricier lab apparatus like centrifuges and a huge bulky item that looked like a house furnace but was probably a mass spectrometer.

"I need to be in a meeting in twenty minutes, so let's start. Show me the reports." Calm looked at the manila envelope in her hands.

"I'll show you this, but first I need to talk to you about a GreenWorld representative who came by our lab. We had some grad students who wanted a project connected to *the Scrubber*, and as soon as they started their analysis, this GW guy comes and starts asking questions and demanding access to their preliminary findings."

Calm seemed to become a little agitated. "What are you trying to say, Ms. Koss? Do you think GreenWorld would interfere with another lab's studies?"

"Well, we didn't think any serious scientist would do that, but then we had a break-in in our lab. Laptops were stolen as well as notebooks. Is this how Green-World works?" Jane tried to sound confident, but she knew her voice was shaking.

Calm broke out into a sweat, "Ms. Koss. Let me show you how we at GreenWorld work." He took her shoulder and guided her silently into the next room. There a garage-door sized panel covered one wall.

"Here, put this on. It's cold in there," he handed her a bulky white jumpsuit that looked like it came from a space station, then took one for himself.

All of Jane's alarm bells were going off as she stepped into the suit. *Don't do what he says. Run!*

Calm was acting like he was on the edge of panic, sweating profusely, muscles twitching and clumsy. He reached over and grabbed a large chrome handle the size of a baseball bat and pulled.

The panel opened to reveal a huge freezer with a cloud of condensation vapor rolling out. "Look at this," he said, pushing her into the freezer.

86

THE EARLY DAWN had brought a stunning beauty to the woods around the Alaska Remote Climate Station. The recently departed storm seemed to have scoured ice crystals from the air and all surfaces; everything appeared unusually bright and in-focus. Soon the remote beauty of the ARCS would be replaced by the noise and congestion of the Anchorage airport.

Dave White glanced back at the log-built ARCS before he boarded the Alaska State Trooper A-Star 350 helicopter. He thought about the meeting he would soon have with Governor Bates-Hardy. The pontoon-equipped aircraft would travel at 150 miles per hour to Anchorage where the governor would demand answers to rumors she had been hearing.

He had been up all night reviewing *the Scrubber*'s performance data. His frustration mounted as he tried to balance scientific integrity with performance goals. And those performance goals were doable with the money in the GreenWorld-embossed envelope directed towards procuring needed equipment. But, the "we only need a little more time" perspective with its associated deceit had been washed away with last night's storm. Dave knew what he had to do. He waited nervously as the powerful helicopter sped towards the meeting he'd demanded and dreaded.

SHIRLEY AHN led Dave White into the formal visitor's office in the Anchorage Governor's Mansion. She had him sit in the uncomfortable chair facing the huge ornately carved desk. He noticed a shiny Dr. Pepper can looking out of place among the Victorian furnishings.

To the side, a hidden door burst open and Governor Bates-Hardy strode out carrying a coffee cup in one hand.

"Professor White, please excuse my abruptness but we're running late. Greg Yeung's media team is waiting in the lobby and they seem really pleased with themselves. I need to meet with them very soon, so please succinctly summarize your report."

"Governor, you know how important our work at ARCS is to Alaska and the world," Dave started.

"Professor White, I'm in a terrible rush. Please just tell me why you asked for this urgent meeting," the governor put her coffee cup down with a loud clunk and stood over White.

"Yes, Ma'am. In order to give the ARCS team more time to solve some of the little surprises, I added secret boosts to *the Scrubber* and manipulated test data. To an outsider, the test data indicated *the Scrubber* was showing the whole system was ineffective. I had to shield my team from that information to keep up morale."

"You faked the data?" The governor's voice was stoic but her knuckles turned white where they gripped the mug.

"No damn it! I didn't fake anything. I just held back a few irregularities. That's not faking data. I don't expect someone outside the scientific community to understand."

"So you were selective in the data you reported?"

"Yes. But unfortunately Governor, the evidence is mounting that those irregularities may in fact represent a trend."

"Don't tell me about irregularities. Just say it, does *the Scrubber* work? Can we burn coal without releasing greenhouse gases?"

Dave felt sweat break out all over his body, "We just need a little more time to optimize the Scrubber's performance. We're right on the brink of a huge break-through."

"So it doesn't work. You just made up the little part where it worked." Leola's voice became soft.

"We weren't just making shit up. I believe in *the Scrubber* and, in general, the data supported our performance expectations." Dave felt a bead of sweat run into his eye. "Governor, you were struggling with the environmental voters. Yeung and Bolton were making significant inroads with those voters looking to accelerate the growth of the Alaska economy. Your timeline was pushing us to minimize the scope of our analysis."

"Why did you choose to tell this to me now, Dave? Why not before I announced that *The Scrubber* would save the world?"

"I thought you would want to know before someone else told you."

"This is not about me, Professor White. Nor is it about you lying to me and the citizens of this great state. This is also about your lie to the people who trusted you and poured their energies toward supporting this project. This is a professional disaster for your team! They are all stained by your actions. You have totally destroyed their credibility and their careers.

"Professor, you seem to have forgotten the theme of one of our first meetings. My tenure in office and my re-election campaign are built primarily on two principles: integrity and transparency. And I have been true to those principles. Until today. Because of the position you put me in, I have spread lies to people all over the world!" The governor's self-control mechanisms strained as her frustration ratcheted up another level.

Dave wanted to provide the governor with a little more detail, but he was certain that this was not the best time. "I apologize for the incomplete data, but the results weren't that bad, Governor. I just thought you would want to know before the press got a hold of the story."

"I want a full report of *the Scrubber* project on my desk in 24 hours. Be sure to include everything you know that you hid from your team." Her voice rose. "You are going to be very sorry you put this hoax on Alaska! Very sorry!"

Leola abruptly left the room wondering what she would do when Yeung was governor.

87

TANGO KNEW HOW to locate, track and kill a target. She knew how to use her assets to best accomplish her other military goals. She had practiced her skills working for many governments and organizations. At this point in her career, she could tell when the party who hired her was becoming unraveled. When this happens, it is time to use the utmost caution so she could extract herself and her assets.

Looking over the new emailed orders from The Boss, she realized again that her boss did not have a military mindset. An escape plan started forming in her mind. It wasn't time to leave yet, but she would need to be ready.

Her foot soldiers would do what they were told. She knew they would sacrifice themselves if she told them it was for their cause. But the new kill target was so significant, it might provoke them to stop and think. She had to wait until shortly before the kill to tell them who they were after.

Then there was the problem of quickly getting her troops to the new locations. They would need to travel several hundred miles very fast and carry some heavy equipment. She decided to make the trip using several floatplanes. The planes could land in a nearby lake. The damn rivers close enough to the Zetros were too shallow to land a floatplane and the wide cobblestone riverbeds too bumpy to land a plane—even with tundra wheels.

Tango made the appropriate arrangements then prepared the weapon she would deploy for the high profile target. Her first choice would be a shoulder-launched, multipurpose, assault weapon.

When she was in Afghanistan, they called it a SMAW. It was deadly within 500 meters, but it was too bulky and, since it looked like a bazooka, it could not be carried around in Anchorage without drawing attention from local police.

From the Zetros armory, she had brought a Russian-made MRO-A. It was basically a disposable metal tube, less than a meter long and weighed only 10 pounds. Tango knew it was deadly accurate within 90 meters, which would be perfect for the location of the target.

What was most important about the weapon was its small rocket, a thermobaric warhead. These amazing warheads killed with a blast wave created when all the oxygen in the large blast radius is consumed by the explosive charge. Buildings and lungs are ruptured and destroyed.

The one in Tango's hand had two separate explosive charges. The first charge was designed to burst open and disperse a toxic cloud of propylene oxide that mixes with the atmospheric oxygen. If this cloud is inhaled, it burns a victim's lungs and causes painful suffocation. When the second charge explodes, it ignites the cloud creating a massive lethal blast wave. The thermobaric weapon works best indoors, and that's where the target would be waiting for her, the Botanical Gardens.

As she finished packing, something felt odd about pulling up and leaving ARCS. The Boss seemed to have lost interest in monitoring the scientists, which Tango had found boring, but necessary. This mission was starting to unravel, and killing this next target might just start a complete cluster fuck.

88

THE CESSNA TOUCHED down lightly in Anchorage and taxied to the general aviation area. Gator made arrangements for fuel, used the bathroom, and checked in at the office of her bush pilot service.

The others stretched, picked up bits of beef jerky and other trash from the floor, and loitered near the plane. Sully noticed Gator emerge from the office. She approached the group with a furniture dolly stacked with boxes and a brown paper bag much like the one that had earlier held a dozen candy bars.

"Get your stuff out of the Cessna, we're taking the Cherokee to Seward." Gator approached a handsome red and white Piper Cherokee and opened a cargo door near the rear. Dylan could see 4 roomy passenger seats covered in pristine white leather. This was going to be a far different flight than the rickety Cessna 172.

"I want Fatso to sit in that rear-facing seat." She looked at Sully then gestured to the place behind the pilot's seat. Gator pointed to Taylor, "I want Skinny sitting next to him." After she had shoved the last box in the cargo hold, she turned to Dylan. "You sit up front. Don't touch anything and keep quiet when I say so."

Soon the plane was loaded up, except Dylan. Gator had him wait outside as she performed the exterior preflight routine and talked to her passengers through the open doors. "This is a 1968 Piper Cherokee 32-300. The 300 stands for horsepower, so you'll notice it's more vigorous than the Cessna. Since I'm hauling 80 pounds of cargo in addition to you guys, we couldn't take the smaller plane."

She slammed the cargo door shut and locked it from the outside. She opened the door on the right side of the plane and stepped on the wing to get into the cockpit. Once there, she slid over to the left side and gestured for Dylan to get in.

"I want it quiet while I do the interior preflight," Gator said. Everyone behaved as she went through her checklist.

Sully tapped Dylan on the shoulder and pointed to the paper bag on the floor. He indicated that Dylan should pass it back. Dylan, unwilling to "touch anything" shook his head.

Gator started giving a rundown of what she was doing, "Now that preflight's done, I turn on the fuel pump." She leaned down to her left and flipped a switch. A whirr could be heard from the engine area.

Sully again gestured to Dylan to pass the bag.

"Then I prime the throttle three times." She pulled a knob in the center of all the controls out and pushed it back three times.

Sully opened a compartment under the cup holders and found a cache of water bottles. He pulled one out, opened it and took a drink. Then offered it to Taylor.

"Put the mixture on full rich," Gator made another adjustment. "And crank the engine with the magneto off to circulate the oil."

"Want one?" Sully showed the water bottle to Dylan.

"Clear the prop!" Gator yelled causing everyone to jump. The prop spun in a jerky way several times. She flipped the magneto on and the engine roared to life.

Dylan wondered if Gator was going to narrate her actions for the whole journey, but as soon as they were airborne, she piloted without comment. From the rear, Dylan could hear Taylor and Sully discussing laser physics, non-biological oxygen production and other processes far beyond his scientific knowledge.

He wondered just why his crude college science project had ended up having such an impact. Dylan speculated that college labs all over the world work on breaking up carbon dioxide. It's probably a standard exercise.

He missed Suzie. He had been guiding a trip when she left on this four-week trip. He wished he had canceled his hunt and taken her to the airport. Dylan wondered if he and Suzie were drifting apart.

Suzie would ask him questions to clarify his thoughts. He wondered what she would ask, then he started composing questions for himself. *Why my design? Surely there were top designers who had come up with a framework that was more modern. Why were we being shot at? Why was Ethan killed? Was Zhang murdered? Why did being hunted by those killers feel so similar to ten years ago when some mercenaries hunted him? Would the soldiers follow Dylan and the rest outside of the Denali area?*

Sully turned to Taylor, "The good news is that our CO_2 dissociation model provides us with an easy way to prove or disprove whether this process is active. The bacteria scrub the CO_2 out of the exhaust gas by combining it with hydrogen gas. All we have to do is to monitor *the Scrubber*'s exit gases hydrogen gas and carbon dioxide gas levels. If both CO_2 and hydrogen gas levels drop, the bacteria are doing their job. If only the CO_2 concentration decreases something's rotten in Denmark."

As Sully and Taylor's discussion again veered into an esoteric range, internally Dylan continued composing questions that Suzie might ask. *What's with Anja? Does she want a relationship with me, or is she using me for her own purposes? If she's using me, can I still find a way to be a dad to her? If she sees that I can help her with the governor, will she warm up to me?*

Gator picked up the paper bag on the floor. Sully stopped talking. How he knew she picked up that bag, Dylan couldn't tell. "Anyone want a snack? I see you found the water bottles." She passed the bag over to Dylan. He peered into the sack and smiled.

"Sully, you pick first." Dylan handed the bag over seat.

Sully looked into the bag and his face fell. "Carrots? You brought baby carrots as a snack? What happened to the candy bars?"

"I'm pregnant and you're diabetic. Neither of us needs the extra sugar. Enjoy the carrots."

Sully pulled out several and gloomily began eating them. He seemed able to make incredibly loud crunching sounds.

Dylan returned to his reverie. *And the governor? Why is she showing interest in me? I'm sure it's to get Anja's campaign contributions.* Dylan thought about the powerful forces involved around *the Scrubber*, big campaign money, big energy concerns, possibly millions in revenue once *the Scrubber* goes commercial and a step towards reversing climate change.

Another loud snapping sound from Sully's carrot crunching brought him out of his ponderings.

"Dylan," Taylor turned in her seat, "Are you sure you have the report on *the Scrubber* at your house? It could clear up a lot of questions."

Sully spoke through a mass of orange carrot bits. "We might find out that *the Scrubber* doesn't work at all, or that it will save mankind."

"My daughter said she'd send it to me. I have no reason to doubt her."

"OK everyone," Gator announced. "Shut up. We're approaching Seward."

Dylan looked over at the massive, sparkling Harding Icefield. At 1,000 square miles, it sends over 40 glaciers spilling downward. Exit Glacier was clearly visible glowing a soft blue as Gator put the Cherokee into a landing pattern. As the wheels touched down near the number 34 runway, Dylan felt a feeling of excitement come over him. Soon he would be home.

Suddenly Dylan longed for his old life in Seward. He wanted the profound sense of peace from living in his cabin-in-the-woods with his beautiful wife.

However, Dylan could not shake off vague, anxious feelings.

89

JANE NEARLY SCREAMED when pushed into the huge freezer. Around her were animal heads and other body parts in clear vacuum containers. Unrecognizable internal organs were tumbled on shelves, and she wondered if they were human parts. Calm pulled the freezer door closed and it latched with a final-sounding click.

"I want to get out of here," Jane said, her voice quivering, but Calm's presence was eager and forceful.

"We can talk here," he said, pointing to the bracelets. "They don't record. Only transmit. The outgoing signals from the bracelet however cannot penetrate the walls of this freezer. Walls are too thick. So security puts sensors and transmitters in here, which I disabled with liquid nitro. We have just a few minutes before they start looking for us."

He's afraid of GreenWorld. He doesn't want them to hear us.

"I don't know if that was a GreenWorld operative who stole your experiments. Before you got here, I got a message from security. They are uncertain that you work at a university lab. GW thinks you might work for some guy named Colonel William Bolton, Greg Yeung's guy."

"That's bullshit. They are trying to manipulate you," she said. *Play into his paranoia. Tell him what he thinks is true.*

"I figured as much. Listen, I'm really trapped here. Don't tell anyone that, but it's true," He hiccupped as if overcome by emotion. She watched him stand there sending out clouds of condensation. "I need to tell someone, so you can tell if they come after me or my family."

"Tell me what? What do you mean?" Jane became alarmed at his discomfort.

"OK, Dave White sent components of *the Scrubber* over here for our labs to test. He gave very detailed instructions about how we should test it and to ignore anything unusual that we notice."

"That's not normal?" she asked.

"You know it's not normal for a lab to be told exactly what to look for but not to pay attention to anything else." His eyes expressed doubt. Maybe he was wondering if she were really running a lab.

"Of course I know that. Go on," she bluffed.

"OK, so we get the equipment and protocols and start the test. And then I get a call from Dr. Neil Carlyle, Dave's boss at University of Oregon. Neil says that he's noticed some really funny things going on with *the Scrubber* project and could I do some other tests."

"Other tests?" she asked.

"Well, not really tests. He wanted me to do a thorough analysis of the design and construction of *the Scrubber*. He said he thought the design included components necessary to build a laser." He looked at her like she should be shocked.

"You're shitting me!" Jane said. *Why the hell would there be a laser?*

"Yeah, right? A laser! As if! If they put a laser in *the Scrubber*, it would use more energy and thus produce more CO_2! Plus, a laser is so old school. There's nothing innovative about using a high-power laser to dissociate CO_2.

"Well I told him I'd look into it, but then I hear that he died in a freak mountain accident." Calm looked close to tears. "He's dead! I'm wondering if they killed him."

"I'm sorry," Jane didn't know what to say.

"Then I get a call from upstairs. You know what I mean, UP STAIRS!" His voice rose as he pointed upwards, "Telling me to disregard Neil's instructions."

"I wonder if Dr. Carlyle was killed to shut him up," Calm looked close to panic. "And I wonder who's next."

90

TANGO AND RENO stashed their Lightning LS-218 electric motorcycles in the bushes about a quarter-mile from the Botanical Gardens. They would hike in and set up their strike.

Tango had the most experience with the MRO-A handheld rocket launcher, so she got to take the kill shot. Reno would have the second shot in case Tango missed, and he would cover their getaway.

They had arrived early. She watched Reno brush his long red hair. The motorcycle ride on the way over had messed it up. She had worn a helmet, but Alaska didn't require riders to use protective gear, so Reno got to have his long red locks wave in the wind.

Tango visualized a torrid, high-energy romp in bed as she admired his well-muscled body. She wanted to test some recon ideas that would certainly get his attention!

If she ever did get him in bed, he'd have to keep his mouth shut.

She didn't tell him who their target was because it might affect his functioning. She told him The Boss decided the target was necessary to take out, so he would follow orders.

On the way over to the Botanical Gardens, Tango found it eerie to ride a nearly silent motorcycle. Their Lightning LS-218s had been fitted with off road tires in case they needed to avoid law enforcement off the streets so there was some tire noise. However the acceleration of the small bikes was so rapid, she knew they could outrun cops. The bikes could top out over 200 mph, and could go from 0 to 60 in less than two seconds. No police car on the planet could come close to that.

After they located the orchid pavilion, Tango planned her shot. Their target would be waiting at a conference table inside the building, but at least one garage-door-like wall would be rolled up. The Botanical Gardens website had said the pavilion would be open until nine. There would probably not be any other visitors around at that time, so the likelihood of collateral damage was slight.

Who the hell builds an orchid pavilion in Alaska? wondered Tango. *It's so fucking cold here in winter, they must spend a fortune heating the place.* After she sent the thermobaric rocket into the building, they wouldn't need to heat it ever again.

91

JANE LOOKED AT THE PANIC in Dr. Calm's face. *He's really close to losing it!*

Calm took her arm and lead her back deeper into the huge freezer. As they walked, banks of overhead lights came on and winked out behind them. Soon they stood before another massive door with a big chrome handle,

"This is the ultra low deep freezer. It stays at -85 degrees Celsius. You need to wear this to enter, and turn on your jacket," he showed her a switch on the sleeve.

Jane felt claustrophobic when she pulled the tight mask over her head. It resembled a full-face motorcycle helmet that sealed around her neck and plugged into her suit. Nearly instantly, she could feel the suit warming up. An orange LED light indicated her visor was heating, and the fog from her breath vanished.

Around her she could see racks of biological samples all barcoded and carefully arranged. "We do tissue analysis and product testing for many different companies. GW distrusts outside labs, so we built this facility and offset the costs of our investigations by taking in work for other entities. Have you seen anything displaying the PTL logo?"

Jane looked around at all the vials and zip bags displaying the recognizable gold PTL badging.

Calm picked up an empty beaker and showed her the PTL logo, "It's us, Partikkel Tek Labs. GW is actually making money off this lab. A lot of money. They don't tell people that PTL is part of GW, so keep it under your hat."

In the very back of the ultra low deep freezer, a secure vault about the size of a washing machine occupied a shelf. A baleful warning sign: *Caution Hazardous Materials. Authorized Personnel Only* glowed next to a sensor. Calm swung his bracelet past the sensor. With a hissing sound, the door opened.

In a drawer in the bottom of the vault Jane could see a label, *Caution. Radioactive Materials. Use proper precautions.* Without any precautions, Calm slid the drawer open. It was full of vacuum bags of human hands. Jane gasped.

"These are from the bodies of workers who died in a radioactive event in Japan. We've finished testing them and found there's no danger of exposure to anyone."

"If you finished the testing, why do you keep them?" Jane's unease at being in the GW freezer was pushing her toward panic. *I wonder if they put something in the air to encourage paranoia?*

"These body parts keep most people away. No one bothers the stuff I hide here. It gives me a measure of privacy in a place where security always knows everything," Calm said as he pulled out a pouch the size of a sandwich. "This is made of rubberized lead, the same stuff as those bibs the x-ray technician puts on you at the dentist's office." Calm put the package in the pocket of his heated jacket. "Let's get out of here."

Jane was only too happy to follow Calm out of the ultra low to the regular freezer. They returned their helmets to the rack and stood under a light. Calm opened the pouch made of rubberized lead and shook out a tiny black piece of plastic smaller than the size of Jane's little finger nail.

"This is a micro-SD card. It has a copy of the report containing *the Scrubber* specs we created in this lab," Calm looked significantly at Jane.

"What's in the specs that's worth hiding?" Jane asked.

"We were ordered not to study or comment on anything outside of the very strict parameters given to us by Professor White. But I left some subtle hints of *the Scrubber*'s problems in this report."

"What kind of hints?"

"Oh, just little things like laser components," Calm smiled.

"Really? Lasers that assist *the Scrubber* in removing carbon dioxide?" Jane just hoped she was using the terms correctly.

"Exactly, some lasers can separate the carbon from the oxygen in CO_2 molecules," said Calm. "It's not just germs that break up CO_2."

"So can the Scrubber actually work?" Jane shivered in her heated suit.

"Of course it can. It's amazingly effective," said Calm. "But it's impractical. Maybe future scrubbers will be developed that do the job efficiently, but I'm guessing we are all going to need to get used to producing less carbon dioxide."

Suddenly a loud *click* sound interrupted Calm. The door to the freezer opened and Jube stood there scowling.

"And that's why it couldn't possibly be anyone from GreenWorld harassing your lab. We are an open book here. If you want to look at our data, you can have it as soon as it's published." Trevor Calm looked at Jube, "Jube, do you need something out of the freezer? I was just showing Dr. Koss our lab and discussing her report."

Jube looked dubious. "Nah, your bracelets just dropped off the system, and we were wondering what happened."

"They dropped off? How strange. Could it be they don't transmit in the freezer?" Calm asked.

"We'll need to put a sensor-transmitter in here," Jube looked around shivering.

"They've done that, but they keep failing," said Calm looking significantly at Jane.

Jane pushed out her hand to Calm and shook his hand feeling the cold micro-SD card being pressed into her palm, "I don't know how we could have been so wrong, Dr. Calm. Thank you for explaining your test protocols to me. You'll need to drop by our lab sometime. I'll return the favor you granted me and give you a tour."

She turned to Jube and kept her voice steady, "Can you show me out?" *I've got to get out of here!*

92

IN THE TINY SEWARD airport, Taylor settled the bill with Gator and over-tipped using the ARCS credit card. They walked to one of the charter service offices where a battered green minivan sporting a hand-painted sign, *Lisa's Taxi Service*, was parked. Dylan greeted Lisa, a round, energetic, elderly woman dressed like an auto mechanic in green long-sleeved overalls.

After pleasantries, the party got into the surprisingly clean van and started off. Sully asked Lisa where they should eat in town. While Lisa, Sully and Taylor discussed the relative merits of the various restaurants, Dylan dug around in Taylor's pack for her phone.

He held it up to her and pointed to it, asking permission to make a call and indicating his phone was dead. Taylor nodded without breaking the conversation about food. After a frustrating five minutes on hold, Dylan got through to Anja in her Anchorage office.

Then Dylan had to put Anja on hold while the party discussed how Lisa would drop them at the restaurant and pick them up in an hour. Soon, Dylan found himself standing in the Harbor View Restaurant parking lot by himself while Taylor and Sully went in to grab a table and order.

"Dylan, what's going on?" Anja demanded.

"We just arrived in Seward. We think *the Scrubber*'s test data was altered. The Scrubber may not work at all," Dylan said.

"Do you know how many people would be ruined if that information got out?" Anja hissed. "Whatever you do, don't tell anyone because there are powerful people who would kill to suppress this information."

Dylan immediately thought of Cheek, that bastard. If *the Scrubber*'s data were intentionally faked, he would suspect Cheek. The governor needed an actual way to burn coal without creating greenhouse gases in order to engage a skeptical constituency, and Cheek was running her campaign.

"I'm just leaving for a meeting with Cheek," said Anja. "Let me find out what he wants before you discuss this with anyone. Am I clear?"

Dylan felt uncomfortable being spoken to like a child. Maybe some of Bolton's imperious manner had rubbed off on his sweet daughter. "Anja, of course we won't discuss it until we have proof. The ARCS scientists with me are certain that the research notes with all *the Scrubber* specs should be able to confirm or negate our suspicions."

"Who else knows where you are right now?" Anja said, rushing the words out quickly.

"No one," said Dylan. "Well, our pilot and a few others unassociated with *the Scrubber* project."

"Be careful. Be very careful. Are you armed?" Anja asked.

"No, but I have some arms at my cabin. I'll pick up the notebook and take it directly to my cabin."

"I think you are in great danger. Those state troopers who tried to kill you may still be after you."

"Anja, we know for sure that one of those guys is dead, and it's very likely the others were buried in the coal mine." Dylan shuddered thinking back to being trapped in the mine. "Anja, I want you to be careful meeting with Cheek. He gives me the creeps."

"I'll be careful. I'll have Roger and Andrew nearby. I got to go. If you want to call me again, use my personal phone number. Very few people have this, and I always answer." Anja ended the call. Dylan stood in the big paved parking lot feeling exposed. He wanted to get to his cabin. He kept telling himself that in the woods, he would feel safe.

ANJA PACKED UP her briefcase in the event she actually got to negotiate with Harold Cheek. She didn't trust him. His plan was probably just to issue edicts. Not only was he disgusting and manipulative, he would say anything to further his aims.

She cleared off her desk and thought back to her first meeting with Cheek, Governor Bates-Hardy and Dave White. It was at that meeting that White presented *the Scrubber* models to Anja and the governor.

Anja remembered Cheek lurking in the background staring at her chest. He was so overt about his gawking. Unconsciously, she buttoned the top button of her blouse.

Anja punched the elevator call button and waited. She remembered how deftly White had answered her questions. His contagious excitement about how *the Scrubber* would slow global warming and how *the Scrubber* 2.0 and 3.0 would be part of every carbon dioxide emitter made by man, actually moved Anja.

White particularly wanted her to know that his team had chosen Dylan's design for *the Scrubber*. It was as if he thought that alone would make her approve a multi-million dollar research grant for his department. His excitement focused more on the future potential of *the Scrubber* rather than its present capabilities.

Anja knew to be on her guard. Since working in the fund-raising areas of non-profit, she'd done her share of manipulating. White would need more than Dylan's scrubber design to get a grant from the Mountain Club.

White had brought only one copy of the research notes with *the Scrubber*'s specs. He told everyone it was highly confidential. Anyone at the table could study it, but it would have to stay with him.

Instantly, Cheek had grabbed it and carried it to another table. As the meeting went on, White noticed that Cheek was photographing some binder pages with his camera.

THE ELEVATOR car arrived with a loud chime, and Anja found herself its only occupant. She found it peculiar to be without her ever-present shadows, Roger and Andrew. As the elevator descended, she thought of White's reaction when he noticed what Cheek was doing with the camera. White looked like he wanted to attack Cheek with tooth and claw. It couldn't have been better. Cheek had fled the room with his camera.

"Since the governor's office now has a copy, my team at Mountain Club would like one too," Anja stated. White had handed over the binder to Anja who noticed all the tests were certified by PTL. It made her smile to see the familiar gold-colored logo of the famous testing lab on the cover.

"Don't lose it or leak the information," Cheek had sneered when he returned to the meeting. "We know how you political types are with secrets."

Anja had fumed. There was no one more political or underhanded than Cheek. And he had told her not to do exactly what he would have done. He would use any means to accomplish his goals.

She wondered what he would have thought if he had known she had sent the binder to Dylan's address in Seward. At no time had she told Dylan the information was confidential. When she had thumbed through the binder, it appeared to be written in a foreign language. It was doubtful that Dylan would bother to look through it. He didn't seem the technical type of guy.

When the elevator doors opened to the garage, Anja decided to take one of the Mountain Club company cars to the appointment with Cheek. At first look, the only available vehicles were large delivery vans. No sense burning all that fuel just to go to the Botanical Gardens.

Then she saw the small car in the corner. It would be just right.

94

TANGO SET THE ROCKET to detonate at 50 meters and made herself comfortable in the prickly Alaska bushes. She watched a group of senior citizens spend an absurd amount of time looking at and discussing orchids. The flowers brought to mind a mission she had in a cloud forest in Peru. The place was awash in orchids ranging in size from a pea to a grapefruit. These idiots were going crazy over such ordinary flowers.

Then it drove her crazy to see several teens from a nearby high school set up at the table and take out notebooks as if to do homework. If they occupied the table, the target might pick another place to wait.

Winds tossed the overhead treetops. A forecasted storm seemed to be coming in early. She and Reno had a small plane waiting just a few miles away on a gravel runway at an uncontrolled airport. If the storm came early, it could interfere with their trip to Seward to meet the rest of their team and complete their list of kills. Bush pilots were often fearless, and only the living ones were cautious.

She thought about telling Reno that he might need to hunt the target if all those people stayed in the orchid pavilion. The target would find a different place for the meeting. Once Reno saw the target through his scope, he might question the orders. She really wanted those people to leave.

Then she saw a Nissan Leaf pull up near the pavilion. She couldn't figure out who would drive an electric car in Alaska. The batteries were grossly inefficient in cold weather. Besides being ugly, it was a poor choice for the climate. It must be someone who was making a personal or political statement by driving such a car. It had to be the target.

Ten minutes before the set time for the meeting, the seniors and teens left the pavilion. After hesitating a long minute, the target got out of the car, walked up and sat down at the picnic table checking a phone.

Tango coldly put the launcher on her shoulder, sighted the target and pulled the trigger. The tube barely bucked as the rocket launched, sizzled across the open space and detonated in the pavilion. The explosion sent a fireball and debris rising thirty feet into the sky. A concussion wave broke out the windows in the cars parked nearby. The target and 700 orchids died nearly instantly.

Tango stashed the tube in the bushes and sent a text to The Boss: *First target down. On the way to Seward.*

95

JUBE ESCORTED JANE from the elevators to the security-check room without any conversation. He gave her the jitters, but she guessed it was his job. She wanted to bolt out of GreenWorld and lock herself in a closet.

"We'll take off your bracelet now. Next time you come to visit, the check-in will be much quicker." He indicated that she should put her hand into a black hole in the wall, then he took a step backward to watch.

Jane felt herself on the verge of a panic attack. She did not want to put her hand into the hole. She knew it was irrational, but something told her they would cut off her hand, or insert a probe under her skin, or microchip her like a pet. The hole took on a profound malevolence that seemed out of proportion to reality.

"I don't want to put my hand into that black hole. Just take off the bracelet with that tool, and I'll go." Jane tried to keep the quiver out of her voice.

"Ms. Koss, this is how we take off the bracelet. Our security protocols do not allow our visitors or employees to see how this is done." Jube sounded impatient.

"OK, well I'll just take if off at home and send it back to you," Jane said.

"You can't remove it yourself without damaging it. Remember you signed our agreement. You promised not to take the bracelet out of the building," he said.

Then he came around the counter and put his hand into the hole. She heard a buzzing sound followed by a metallic *click*. He removed his hand now without a bracelet. "Now you try it."

Jane forced herself to put her hand in the hole. The sounds triggered her panic. She jerked her hand out of the hole scratching her wrist on something sharp. A thin red line near her thumb produced a few drops of blood.

"Why did you do that? Just wait until it's off before removing your hand. It doesn't hurt." Jube handed her a tissue.

"Silly me," Jane said. "My things?" she pointed to a locker.

"Your things are in a different locker out in the lobby. Please visit us again sometime." Jube opened the door to the lobby, and she stepped out.

As she collected her purse, Jane wondered if it and its contents were now filled with tracking devices or worse. She waved at Evelyn on her way out trying to look casual.

She had to get away from this creepy place. Not only that, she had to inform her boss about what she had learned.

Her Suzuki X-90 crunched down the gravel drive and out of the huge gate. A feeling of relief flooded over her. She wanted to take a shower when she got home. Jane became aware of a need to throw out her purse and replace everything in it.

As she approached Campbell Airstrip Airport on the outskirts of Anchorage, Jane made a mental list of what she would tell Bolton: *the Scrubber may be a fraud, the*

Scrubber design may include a CO_2 splitting laser, evidence of fraud may be concealed, GreenWorld might have had something to do with the death of an Oregon professor, and we have a new confidential source in GreenWorld. And . . . I have proof of all this on a microSD drive.

Her thoughts were interrupted by a bright flash that lit up the trees on the left side of the road. Jane looked to the right to see a huge fireball rise up into the evening sky. A thought crossed her mind, *A plane must have crashed at Campbell Airstrip.*

96

THE BOSS NEEDED a break. The right breather might be a walk to an ice cream shop about a mile away. The streets of Anchorage would be a place to walk off anxiety. Looking at the reflections of city murals in the Dena'ina Convention Center windows, The Boss thought about how complex everything had gotten.

It had been easy to get *the Scrubber* project going. Given enough money, The Boss was able to find a prestigious university to develop *the Scrubber*, put a stamp of approval on it, then send it along to GreenWorld for further testing. The problem had been Neil Carlyle. If he wasn't so meddling, he'd be alive right now. But he had to poke his nose into the project and find out about the laser.

It caused The Boss great discomfort, but Carlyle had to die. To keep authorities from looking too close at his death, it needed to appear to be accidental. The Boss had provided a large anonymous fund to care for Carlyle's family and no one seemed to be the wiser.

A door on the convention center opened and a stream of conventioneers streamed out. It had to be the Reefer

Rendezvous, a huge annual convention of cannabis entrepreneurs. Judging by the funny marijuana hats, tee shirts and shopping bags, the convention was in full swing. That probably meant the line for ice cream would be an hour long. The Boss decided to just get a cup of coffee, really good coffee.

It made The Boss feel sad to think about the others who had to die. Zhang discovered how *the Scrubber*'s data were being manipulated, so he had to go. Then the ARCS away team went where they shouldn't be. All they had to do was to stay put. Ethan died quick with the shot to the throat.

Then The Boss felt an emotional jolt thinking about the loyal trooper, who had given his life to try and eliminate the rest of the ARCS team. Such a loss of a loyal soldier! The Boss would need to care for another family.

Arriving at the ice cream store, with its line going nearly halfway down the block, The Boss turned instead into a gourmet coffee shop to order something special.

Waiting for the order, the Boss thought about the most recent kill, a prominent Alaskan figure at the Botanical Gardens. There would certainly be a thorough investigation, but maybe some terrorist group would take responsibility. That would make it easy.

As The Boss sipped the fragrant coffee, the idea of killing Taylor, Sully and Dylan came to mind. It was time for these troublemakers to go, but it was still regretful in a way.

Feeling sad about all the necessary deaths, it was a somber walk back. *None of it needed to happen, but the goals behind the Scrubber project were bigger than the small people who got in the way.*

97

AS DYLAN SAT in The Harbor View Restaurant and watched the clouds skirt around the emerald snow-sprinkled mountains surrounding Resurrection Bay, he thought he should be filled with a calm, relaxed feeling. In the foreground, the harbor cruise boats came in, expertly docking and disgorging passengers. Soon the bar part of the restaurant would fill with tourists and boat crews.

It would get loud and jolly in the comfortable evening. This was Seward. This was home. But uneasiness kept intruding on the peace. It was as if a tiny voice inside was telling him that things weren't right at his forest home.

Sully urged everyone to try a second bottle of wine he had ordered to go with his pricy butter-braised-halibut-topped-with-candied-hazelnuts.

Dylan noticed how pretty Taylor looked in the evening light and missed Suzie. If Suzie were here he would put his hand on her cheek and brush back her hair. He would draw her face to him and gently kiss her cheek, nose and ear.

Taylor boldly looked at him, and he quickly looked away feeling guilty. *What's wrong with me?*

Later, just as Sully was digging into a huge dome of ice cream covered in golden-brown meringue, Lisa came in from the taxi waiting area to say it had been an hour and where did they want to go next? She sat down to help Sully with the enormous dessert and gossip. Everyone caught up on who was piloting which boat, the latest with her knitting group and the price of fish.

As Dylan let the drone of Lisa's voice soothe him, he suddenly felt his heart race to full speed as two fit-looking men with tool belts and black jeans took a seat at a booth to his left. He studied the men until he realized he'd seen them before working on the docks. His heart rate began to return to normal. *What is happening to me?*

When the bill came, Sully realized he didn't have his wallet, so Dylan and Taylor had to pay for his dinner.

Waiting for the waitress to return, Dylan jumped at the pop of a wine cork being pulled at another table.

This, he realized was a type of hyper-vigilance he'd experienced years ago after a group of deadly mercenaries wrongly targeted him and his friends. His doctor had told him at that time that he was experiencing posttraumatic-stress-disorder, PTSD. *Is it coming back? Is that why I'm jumpy and nervous.*

Dylan knew what to do. His doctor had prescribed cognitive behavioral therapy, a way of thinking where Dylan would identify the bad thoughts and replace them with less distressing images. Sometimes, the bad thoughts came too suddenly for him to fix, like the wine cork sound. He wondered if he needed more therapy.

Everything around Dylan seemed normal as Lisa drove them over to Yukon Dave's Pub where Suzie always parked her pickup when she was out of town. From there, they collected Dylan's mail—which included the ARCS notebook sent from Anja with *the Scrubber*'s specs, some supplies at the grocery store and headed out on the narrow gravel road to Dylan's cabin. Crunched in the bench seat of the pickup, Sully chattered away as he thumbed through the 3-inch thick binder with *the Scrubber*'s specs.

As he approached his home, Dylan thought he should feel relaxed and calm, but an unexplainable edginess seemed to come over him. The beautiful spruce trees with their scaly gray bark should have offered him comfort, but they seemed somehow malevolent, as if they were hiding bad guys with guns.

Dusk was deepening when they arrived at the cabin. Taylor immediately loved the romance of his beautiful, hand-built cabin. The wraparound porch, the large stone foundation, and the fire-engine-red doors and shutters charmed her and Sully. Sully liked the hand-split shake roof, and Taylor gushed over the little hearts cut into the glossy shutters.

It surprised Dylan that he didn't feel the normal sense of pride at looking at his pretty cabin perched on the edge of a cliff with a stunning view of Resurrection Bay. The rugged mountains surrounding the bay were bathed in warm, orange sunset light, but Dylan was unmoved.

What's wrong? Dylan thought. *Are past memories haunting me or is it something different? The woods around my home are making me anxious.*

Once inside, Dylan found himself closing all the window coverings and checking locks, while Taylor made a fire in the floor-to-ceiling stone fireplace, and Sully put the food away. Besides the glossy log walls, the principle decoration was a slender Hawaiian canoe hanging from the vaulted ceiling.

"Suzie is Hawaiian and that canoe is an antique from her village," Dylan explained as they admired the watercraft. "It's past its prime, so now it's a decoration."

"How does a handyman afford a nice little cabin in the woods like this?" Sully asked.

"I built this myself, with a little bartered help from my friends. Mostly I used salvaged parts. It's on Forest Service land on a 99-year lease. The land goes back to the government when the lease expires," explained Dylan.

Sully looked up at the evil-looking leg-hold bear traps on the wall, "Whoa! I bet you catch a lot of bears with those."

"Those traps are antiques and illegal since they cause the animal so much pain. I would never use anything so inhumane on a noble animal like a bear," said Dylan.

Sully opened a large sports drink bottle and settled in a dark leather overstuffed chair with the notebook. He looked ready for a long study session.

Sully looked up. "Hey, this notebook was ordered and paid for by Dr. Neil Carlyle from the University of

Oregon. Wasn't he the guy who died in a freak avalanche on Mt. Hood?"

From the blank looks around the room, it appeared no one had heard of Dr. Carlyle or his death.

Dylan found himself going window to window peering out into the darkness.

"Dylan, what's wrong? You act like you expect an invasion," Taylor said.

"I don't know. Something is making me feel very uneasy. The Ahtna natives speak of their deep, spiritual and loving bond with the land. Sometimes I believe I can feel a bond with the woods. Normally I find the woods very calming, but now," Dylan searched for the best words, "There is a sense of turbulence. I don't know, but maybe my PTSD is returning." Dylan wrung his hands.

"It's probably PTSD. I know of no studies documenting a human-woodland life energetic link," stated Sully without looking up from his notebook.

"I know how to deal with PTSD," said Dylan. "Look, if those guys followed us here somehow, we need to be ready. Are either of you familiar with guns?"

At the mention of guns, Sully looked up, "Not me. Although I'm pretty good with the more sophisticated pool toys."

"I can shoot," Taylor said. "When I was a kid, we'd shoot 22s at small targets when we visited my grandparents. Grandpa called it *plinking*."

Dylan approached a gun safe. "Come over here. You too, Sully." After punching in a code, Dylan opened up the thick metal door. "All these guns are used in my guiding business. There's no military weaponry here, but these are deadly."

He removed a small pistol, "This is Suzie's feral hog pistol. She hunted hogs with her dad in Hawaii and used this when the dogs had cornered the pig." Dylan verified it wasn't loaded and handed it to Taylor. "It's a Beretta Px4 .40 caliber in polymer, so it's pretty light."

As Taylor familiarized herself with the weapon, Dylan went on. "Fully loaded, you have 17 plus one rounds. The barrel rises as you fire it, so fire in two or three round bursts then re-aim. Also, it kicks much more than your grandpa's 22s, so hold it firmly."

Taylor started to look a bit pale, "This is too much for me. At Grandpa's house we just had small revolvers with little bullets."

Dylan's voice sounded gentle, "Taylor, if those killers come back, bear spray isn't going to stop them this time. Maybe this won't either, but please put on this holster and keep the gun on you until we know we're safe."

"OK, Sully. If you're good with pool toys, this is your weapon." Dylan handed him a Remington 870 Super Magnum shotgun with a matte black metal finish. "We're going to load this with magnum buckshot. Just point, shoot, pump and repeat."

Over the next hour, Dylan gave Sully and Taylor more information about how to use their weapons, but by the end of the training period, he had serious doubts that Sully would be able to effectively use the shotgun.

Sully listened to Dylan with the same amount of attentiveness as if Dylan had been a flight attendant and was explaining the exit row seats. Dylan considered locking the shotgun up again but Sully gamely put it within easy reach and again opened the ARCS notebook and started reading.

Dylan added a log to the fire filling the warming cabin with comforting heat and yellow firelight as the rain and wind lashed the forest.

Sully stretched and farted loudly. "Hey, excuse me. I wonder where the energy usage information is buried in this report." Sully looked at the front of the binder, "This was put together by PTL. They do good work."

Sully was unaware that the results in the PTL binder were about to upend his world.

98

IT SHOULD HAVE BEEN a peaceful moment for Dylan. Sully and Taylor sat with the binder between them and quarreled softly about carbon dioxide dissociation, the fire crackled in a friendly way and a mug of fragrant tea steamed at his elbow.

Something was putting him on edge and triggering hyper vigilance. For the third time he checked his own weapon of choice, a lever-action Marlin Model 1895 chambered for the .458 rounds. It had only a 4-round tubular magazine, but the woods around the cabin were dense and this was a short-barreled rifle that had devastating stopping power within 100 yards. He wouldn't need more his Marlin if it got violent in the woods.

"Hey, how come you get a cowboy gun?" Sully whined seeing the lever-action rifle in Dylan's hands.

"This is what I use when guiding grizzly bear hunts. It only has four shots, but if you can't stop a charging 700-pound bear in four shots, then he'll get you anyway."

"Are there grizzlies around here?" Taylor asked with an uneasy rise in her voice.

"Not here, but there are plenty of black bears. They are much smaller and will probably run away if you just fire a round in the air," Dylan said. "However, if a black bear attacks, don't play dead, fight back."

Sully quickly lost interest in the gun and game discussion. He turned to continue his study of the binder. Taylor yawned.

"One of us should stay awake at all times," Dylan said. "Taylor, why don't you take the first nap? I'll wake you in four hours."

She didn't protest, but went upstairs and found the bedroom.

"I'm going outside to check the perimeter," said Dylan. "If you hear a shot, grab your shotgun and be ready to shoot whomever comes through that door."

"Christ, Dylan," said Sully. "You're so melodramatic." Sully opened another sports drink.

As Dylan pulled on his darkest jacket, Sully suddenly stood up, the notebook containing the data falling on the floor. He yelled spraying green beverage all over the place, "Taylor! Taylor! Guess what I found?" When no answer came, he raised his voice, "Taylor! Guess what I found in the specs?"

From upstairs a sleepy voice said something.

"I was studying the data for the exit gas after it had passed through the system during the PTL study. And guess what? Hydrogen gas!"

As if launched from the bedroom, Taylor, dressed in one of Dylan's long tee shirts, leaned over the balcony. "What?"

"Tons of hydrogen gas!" Sully yelled.

"You found hydrogen gas?" Taylor seemed wide-awake despite the chaotic state of her hair.

Dylan's face showed his surprise. "That makes no sense! Even in my basic design the bacteria converted CO_2 plus hydrogen gas into alcohol. There should have been less hydrogen gas."

Sully put his hand on Dylan's shoulder. "Correct, my boy. The absence of CO_2 means very low levels of hydrogen gas present, if any, after passing through the bacteria-coated membrane! The PTL research study seems to indicate some 'unresolved' . . . problems."

"So we have yet another metric indicating that the bacteria were not growing as expected in original system?" asked Taylor still standing at the balcony.

Dylan looked up and could see her white panties from his vantage point. He quickly looked away. He tried to tell himself that he wasn't turned on. Maybe being so close to Taylor was upsetting him.

"I'm guessing there is also a reservoir of alcohol sequestered somewhere in the "L" technology in the ARCS lab. That reservoir would explain Zhang's earlier observations."

"Could it be a conspiracy?" asked Taylor, stunned by the idea.

By now, Taylor and Sully had put the PTL notebook on the kitchen table and were studying it. Dylan noticed that Taylor wasn't wearing a bra under her tee shirt and quickly forced himself to look at the notebook.

"Holy shit! Look at this!" Taylor pointed to a scrambled mess of numbers and symbols that looked like they were taken off a huge blackboard in Einstein's lab.

"Oh my God!" Sully yelled.

"Now what?" asked Dylan.

"Look at the buildup of alcohol in the ARCS membrane readings! That could be what's inhibiting bacterial growth! If we had an adequate way to remove the alcohol, we should be able to improve bacterial growth and *the Scrubber*'s performance at the ARCS lab."

"So what you are saying, Sully, is that the current performance data was faked but there may be a way to get *the Scrubber* system to meet expectations."

Taylor turned to Dylan, her breasts just touching his arm. She seemed completely unaware of the contact. "Dylan, based on Zhang's observation, the CO_2 data from the recorder and now the PTL performance evaluation, the current scrubber design is not working. Someone has altered the ARCS data and the design prototype specifications to give the appearance of acceptable performance. We need to talk to Dave White. He needs to know this right away. Our whole project is being sabotaged."

"Could it be that he already knows this. Someone had to put an alcohol reservoir in *the Scrubber*," said Sully. He grabbed some pages out of the notebook "The backs of these pages are blank. I'm writing up a report to send to Dave and Conrad. The ARCS team needs to know this stuff." Sully squared up to write his report.

"That report will rock the world," said Taylor. "And especially Professor Dave White."

"Actually, he must already know. There's no way he couldn't know this stuff," said Sully, glancing up from the table, his expression grave.

"We need to get this information out. Anja needs to know first," said Dylan. He picked up his Marlin 1895, his cell phone and started for the door.

"Where are you going?" asked Taylor.

"Out to get a cell signal. If there are killers out there, I hope they don't have night vision."

99

THE NETSVETOV HOTEL still sparkled like a diamond through a foggy Anchorage night as Jane entered the glittering lobby and approached the elevator.

The key card snapped smoothly through the reader. Jane would soon be in the office suite. She had to tell Bolton what Dr. Calm had said and show Bolton the documents on the jump drive. It would get Yeung's campaign back on track.

At this time of night, the offices were deserted. Dimmed lights flashed to daylight intensity as the motion detector sensed Jane's presence. Passive electronic office equipment blinked and flashed as if monitoring the office environment for any security breaches or unauthorized activities.

Jane shook her head thinking of what she would tell Bolton. The Alaska Climate Research Station had big secrets. *I can't believe the ARCS scientists lied about the Scrubber and the scrubbing technology performance data.*

As she dug through her desk looking for the card-reader that would allow her to download the contents of the microSD drive, Jane began to plan her presentation to Bolton and picture his joyous reaction. She plugged the drive that Calm had given her into her computer.

She scanned through the information and noticed considerable background information on Dylan Baker's original work and the latest prototypes. Bolton would certainly want to see this background information after hearing about Jane's conversation with Dr. Calm. She tried to imagine the Colonel's reaction to all this information. She would be his hero. Her future path was set.

Her phone interrupted Jane's reverie. She noticed Shannon's name on the caller ID.

"Hi Shannon."

"Where are you?"

"I'm in the office suite."

"What did you find out at that lab?"

Jane's thoughts whirled: *Uh, where to begin . . . the governor's impressive pro-environmental campaign is based on deceit. The governor's campaign is in the toilet.*

Jane needed to talk to Bolton first. She couldn't say anything to Shannon until after that conversation. "I'll give you all of the details as soon as I speak with the Colonel. I thought he'd be in the office."

"Jane, you know what you were saying about Bolton's shaky financial dealings?"

"You didn't tell anyone. Right?"

"Of course not, but you need to know. I just caught a teaser for tonight's evening news. Apparently a federal investigation has collected evidence of financial fraud associated with one of the gubernatorial candidates. Based on what you've told me, we might know who is being investigated. It appears that your boss might be trouble. Jane, he could go to jail."

Jane considered this latest information from Shannon. "Bolton has done an excellent job of juggling complex financial deals and vetting the sources of the Super Pac's assets. He's a master at this stuff. I doubt he would leave any evidence lying around. They are looking at the wrong guy."

"I don't know anything more. I'll try to catch the news later tonight and get back to you," said Shannon.

Jane's uneasiness grew. "The Colonel was very clear when he told me he would never commit fraud. I'm sure he's terrified of jail." Perhaps the Colonel is taking far more legal risks than Jane had assumed.

"But you told me yourself that the Colonel is worried about finances."

Jane calmed herself with a deep breath. Her mind was still for a fraction of a second before Harold Cheek's insider information filled her thoughts. Cheek had said that the Colonel's funds are way down and Yeung's campaign is in the red. *Bolton has done it again; he's committing fraud.*

"I'm sure the feds would appreciate access to any of the information you have collected, Jane."

"Shannon, I'll call you back."

"Are you OK, Jane? I'm worried about you."

"I'm struggling with credibility issues. I have two unreliable sources and need to decide which is more trustworthy." *Cheek or Bolton, which is the biggest liar?* "This has nothing to do with you, Shannon. Talk to you later."

Jane punched in Bolton's number. The call rolled over to a receptionist dedicated to this "special" line. Bolton felt it was important that his office was available 24/7 to his high value colleagues and clients.

"William Bolton's office. May I help you?" Jane recognized Elizabeth's confident voice.

"Elizabeth. It's Jane. Are you at home?"

"Oh, Jane. Yes, I'm still on the clock for another hour. How are you tonight?"

"A little frazzled. I need to speak with the Colonel."

"It is rather late . . ."

"I know, Elizabeth. This is important. I know he is still awake. Please."

"OK. But only because Colonel Bolton was upset that you recorded his so-called, secret meeting. He was spitting nails when he saw the video and noticed his CFO was apparently also making a recording of the meeting."

"Colonel Bolton knows I recorded his meeting?"

"Yeah. The Colonel knows everything. He told me to erase that recording."

"And did you?" Jane asked.

"Of course I did. But, I do keep a backup of every-thing on the server in the closet. He changes his mind on his orders sometimes after looking at his number charts."

"You have a copy in the closet?" Jane thought about the locked closet near Elizabeth's desk.

"Yes, but I'll wipe it tomorrow morning. If he hasn't changed his mind by then, it's safe to erase."

"When did you talk to him about this?" Jane asked.

"He called up from some fancy restaurant maybe 30 minutes ago. Why?"

"No reason. I am just surprised he mentioned this to you."

"Whatever. I'll put you through."

"No, wait! Actually, let's not disturb his meal. Let's let it wait."

"First it's a rush, then it's wait? Aren't we flighty tonight? He'll probably be heading back to his penthouse or the office within the hour. Talk to him then. Good night, Jane."

"Thanks, Elizabeth."

Jane stood motionless for a few moments and then walked over to the picture window. *He knew! Bolton knew I had recorded his secret meeting, and he didn't say anything to me. Why? That bastard has been manipulating me!*

Jane thought about what she should do next. If Bolton had been manipulating her all along, he might be in very serious legal trouble and planning to use her as a scapegoat.

That recording might be the only thing that would save her if Bolton decided to set her up and attempt to shift any blame onto her. As she thought about it, the situation became clear. Bolton was keeping her off guard so he could shift the blame for any illegal actions by the Super PAC onto her. *That bastard!*

She had an hour before he came back.

100

DYLAN LITERALLY RAN out the front door, jogged sharply to his left and disappeared into the woods leaving the cheerful yellow lights of his cabin behind. If anyone was watching his home, Dylan did not want to make that first shot an easy one.

Once deep in the dark woods, Dylan expected to relax. It wasn't working. The familiar woods did not carry its usual calming effect, and he could feel and hear the uneasy animals rustling the leaves. Something was wrong.

Jogging down a trail, he knew he could get a cell signal up on the rocks he had climbed just a couple of weeks ago. Bolton had called him and somehow knew he had been in touch with Anja. Bolton had wanted to discuss his college CO_2 disassociation project. *I wonder if he knew it was a fake?* The Colonel had ways of collecting valuable information. It was probably how he had become so wealthy.

Somewhere in a deep shadow, Dylan thought he heard an unfamiliar sound. As he lightly ran, he had been aware of a porcupine lumbering through the devil's club bushes and an owl calling a mate, but this noise was different.

Dropping to his belly and taking the safety off his rifle, he lay still and became part of the darkness. Whatever had made that noise, seemed to be gone now. Years ago, Dylan had felt he was haunted by the ghosts of people who had died due to his negligence as a mountain guide on Denali. *I hope the ghosts are not coming back to torment me.*

Dylan crossed a creek where he could hear the plop and splash of salmon defending their nests. The familiar sound helped still Dylan's concerns. Looking down to the dark waters of Resurrection Bay, Dylan could see a tiny beach campfire on the opposite shore.

At that moment, he wished he and Suzie were camping on a remote island on the bay. They would eat fresh-caught salmon and wild forest greens, then make love near a cheerful fire. And they would be far from scrubbers, governors, and killers with guns. It seemed so long ago that he was last with Suzie.

Soon, Dylan found himself at the base of the High Rocks cliff where he had spoken to Bolton. He checked the phone. It showed a weak signal that seemed to fade in and out. Slinging the rifle over his shoulder, he gracefully climbed the cliff to where he knew he'd get a strong signal.

Relatively speaking, the lower parts of High Rocks cliff was a basic scramble not requiring ropes or advanced equipment. On a wide ledge, he passed a room-sized open cave that offered a view of Resurrection Bay, and then the rock became steep. As Dylan climbed, his mind left all the disquieting thoughts and focused on his body and the stone.

Once high above the forest, the phone showed enough signal to place a call to Anja's personal phone. Dylan felt happy that he would soon talk to his daughter, but regretted giving her bad news.

Anja's phone went straight to voicemail. *Why would her phone be off? She said she always answers it. Is she OK?*

Dylan knew he shouldn't worry. He thought about how she might react when she found out *the Scrubber* would not help the environment. He wondered if she would ever again show interest in talking or meeting with him now that his scrubber was nothing. Dylan hoped he hadn't lost both the scrubber and his daughter.

After several more attempts at calling her, Dylan called the Mountain Club operator. After navigating a complex phone menu and waiting on hold for several minutes, Dylan got someone who explained that Anja was unavailable. She had left the building some time ago and was in a meeting.

Dylan called her private number once more and left a message telling her the short version of what was wrong with *the Scrubber* and to call him. He stated that he was home and would leave the cabin from time to time to check for messages.

Dylan considered calling Dave White, but decided that it would be better if Sully or Taylor did that. Dave would want to ask lots of questions that Dylan might not know how to answer.

Although he figured she was still off the grid, he checked his email to see if Suzie had sent him a message. *How many times had he checked for a message from Suzie?* Maybe she was unsafe. His worrying was making him so nervous. He had felt her drifting away from him for a long time.

Dylan's heart beat fast when he saw he had an email message from Suzie. How he longed to hear from her during this unsettling time! But her message was even more disorienting than anything else going on. He reread the message aloud to make sure he understood the meaning correctly, a breakup email!

Dylan's mind went blank, as if he'd just undergone electroshock therapy. He didn't remember climbing down the rock face, but found himself walking through his woods as if asleep.

Dazed, he walked through his dark forest trying to process the message. *I lost my Suzie.* Dylan did not know that he had lost more than that. He would never again get a call from Anja.

101

JANE SWIFTLY APPROACHED the door to enter the office suite. As she neared the entrance, her gait slowed. She'd walked the brief hallway from elevator to the entrance hundreds of times, but for some reason this was different. An aura of dread seemed to emanate from the door, as if something terrible waited for her on the other side.

Jane struggled with her situation. *I value integrity!* Jane now found her actions at odds with one of her most valued personal principles.

Something made her pause outside the office suite door, but she wasn't sure what. She knew Bolton would have a record of all key card swipes and maybe even get text alerts when someone entered the office. If he was not out to dinner but was upstairs in his residence suite, he could be in the office in just a few minutes.

She would have to work quickly. She needed to get into the server in the closet, the one where Elizabeth temporarily cached files Bolton ordered destroyed, and download the video file proving Bolton's corruption. If she had a copy of the video, he would not try to implicate her in his fraud, but he might still try to ruin her career.

The sense of dread morphed into a feeling of danger as she pondered her course of action once she entered the office. Her career was in danger, but she felt a solid physical fear as well. *Would Bolton physically hurt me? Would some of his horrible business associates attack me once they knew I had the video?*

Thinking that the video of the secret meeting might be the only thing that could protect her against any legal charges resulting from Bolton's activities, Jane ran her card through the reader, waited for the buzzing sound from within the office, opened the door, and stepped in.

The room felt like something dark, filled with negative energy. This wasn't logical since just a few minutes ago, Jane had spoken to Elizabeth on her desk phone right outside the door to Bolton's office. The place felt empty. But the room also gave her the feeling that no one worked here anymore.

Since the server was not connected to the internet, Jane knew she had to directly plug her laptop into it to access the archived files and copy the video file. She approached her own office door, swiped her card, and entered her office. Once inside, she unplugged the power cable, keyboard and mouse from her laptop and carried it into the outer office used by Elizabeth.

A moment later, Jane stood in front of the server closet scowling. Someone had put a fresh lock on the door. It looked like a crude hasp and a small but sturdy padlock had been inexpertly installed on the outside of the door. The installation looked sloppy, but there was no way Jane was going to get into the closet without a key or an ax.

At the bottom of the door, a chrome mesh screen offered an opening barely big enough for a kitten to squeeze through. She thought about removing the pins from the hinges and lifting off the door. That was her trick when her playful brothers locked her in the bathroom. But this door, like the rest of the sleek office chrome-and-glass look, was designed to have the hinges hidden inside the closet.

The key to the padlock was probably in Elizabeth's desk, but her desk was locked. Jane rattled Elizabeth's desk drawers. The noise made her jump. She wished she knew how to pick a lock. The little kitchen had simple tools and a set of fancy steel skewers for serving appetizers. Maybe they could be made into lock picks.

Jane burned with frustration. How could she learn to pick a lock in the few minutes she had before Bolton would probably show up? She looked down at her computer.

In college, she had repaired the dryer in her apartment by watching a how-to video on the web. There had to be a *how-to-pick-locks* video somewhere. She opened her laptop and entered a search for picking desk locks. What came up was *how to pick padlocks* and *how to make lock picks from paper clips.*

What took just seconds on the video seemed to take hours. However, soon, she was poking the padlock with her homemade picks. Like magic, the lock popped open.

She stepped into the large closet with her laptop and plugged it into the small, tower-shaped server. An image appeared on her laptop screen. It was an icon of the server and a keypad. The server required her to enter a code before she could access it. *What could the code be? Bolton's birthdate? Elizabeth's birthdate?* There thousands of possibilities.

When Jane had been hired by Bolton to run the Super Pac, he gave her an email account and a 6-digit numerical password: 000002. Extrapolating to her boss, she entered 000001 into the keypad. An error message appeared: *Incorrect password. You have two tries left before the server will be erased.*

Jane did a quick online search for the numerology code for *safe. Wait! Elizabeth's desk has that card with Bolton's numerology codes.* Jane located the numbers for each letter in the word 'safe', added them together and entered that number into the keypad. An error message appeared: *Incorrect password. You have one try left before the server will be erased.*

Jane wondered if this would be the time to leave the office. She didn't want to destroy a whole server full of data by entering wrong passwords. However, maybe only files that needed to be erased were stored on this server. If that were the case, the Super Pac would not be damaged by the loss of documents. She decided that the worst that could happen was the proof of Bolton's illegal actions would be erased, and if she left now, Elizabeth would erase the proof in the morning. She had nothing to lose.

Outside the office, the elevator make a *ding* sound that made Jane jump. Quickly she pulled all her mess inside the closet. As she closed the door, she noticed a poorly hidden security camera above Elizabeth's desk. This was probably how Bolton monitored his secretary, but it would also be how he would see what she was doing.

She jumped again when a buzzing indicated someone had swiped his or her card. *Is it Bolton?*

By putting her cheek on the closet floor, Jane could see into the outer office through the small vent screen at the bottom of the door. She smelled dust, and saw Elizabeth's feet cross the reception area and approach her desk. Jane heard keys rattle and a drawer open. It was obvious Elizabeth had forgotten something and had come back to get it. The snap of an eyeglass case closing made Jane jump again. Probably Elizabeth had forgotten her dark glasses.

Then she saw Elizabeth's feet approach the server closet. Elizabeth's sensible shoes were right outside the door! *She's noticed the padlock is open. How am I going to explain hiding in a closet?*

A sharp *click* told Jane the padlock had been relocked. Moments later, the door to the office closed as Elizabeth turned off the lights not tied to motion sensors and left the room. Out in the hallway, a *ding* told her the elevator had stopped at the office floor to take Elizabeth away.

Jane heard a sob escape her. Jane was trapped in an office closet with the padlock on the outside and, undoubtedly, Bolton was on the way.

102

DYLAN TOOK TWO more trips out into a raging rainstorm to call Anja's private number but got no response. Despite a growing sense of uneasiness and the emptiness invoked by Suzie's email, he knew he needed to get some sleep in order to function well. When it was his turn to get a four-hour nap, he took advantage of it. However it seemed like he had just laid his head on the pillow when he heard the crows making a ruckus outside.

Maybe someone is coming? Could it be Anja? Could it be some killers?

Dylan picked up his rifle and checked his wall clock; he had been asleep for his four hours. It was time to relieve Sully downstairs. He also wanted to check out the woods to see if the crows were signaling the intrusion of a stranger or just quarrelling among themselves.

Sully slept on the couch with an empty Oreos wrapper on his chest, and Dylan's phone in his lap. One arm was wrapped around a red, hand-crank am/fm radio. Dylan did not hear the crows just then, but did hear a random tapping sound as the wind blew tree branches against his roof. He wondered why the wind didn't wake him.

"Hey Sully, how about making some coffee for us?" Dylan nudged Sully's leg. "And wind up that radio, I think a storm is coming, and I want to hear a weather report."

Sully sat up spilling crumbs onto the floor and looking around with bleary eyes, "Hey, I wasn't asleep. I was watching out for bad guys. Mind if I keep your phone? I used it to take pictures of my report."

"I need it, so be quick. I'm going out into the rain to try and call Anja. If I can't get her, we'll need to decide whom to call next. It's probably a good idea to wake up Taylor and talk it over." Dylan looked around grimly at the terrific mess Sully had managed to make in his formerly tidy home.

"I need your password to get to the photos on your phone. I didn't need the password to take the pictures." Sully looked at Dylan expectedly.

Dylan gave Sully the password and made a mental note to change it later. He waited while Sully verified that the photos of his report were adequate. "The pictures are fine. Hey your phone is almost dead. I would have charged it, but I didn't know if you liked it with a full battery."

Dylan grabbed his phone and left the cabin. Jogging through the cold, wet woods in the darkness, Dylan became aware of the night creatures settling in to sleep through the day and take refuge from an approaching storm. Crossing the creek turned out to be much more difficult this time. The rocks he had hopped across were under water and the banks were close to overflowing.

Being surrounded by high mountain peaks, Dylan knew it would be hours before sunlight hit the bay turning the deep pewter waters to blue. However, when he got a glimpse of the bay, the surface looked like it was boiling with white caps. The fiercest storms seemed to blow in from the south and flow up to the Harding Icefield just a mile from his cabin. It had happened in 2012 with disastrous consequences, and this looked like one of those.

Once atop the cliff, Dylan struggled to listen to the phone in the gusting winds and driving rain. Anja's phone once again went right to voice mail. Fighting a powerful sense of dread, Dylan began the return jog to his home. Something was wrong, very wrong.

103

TANGO HELD ONTO the seat as the 38-foot Interceptor's twin screws dug into the mad surf of Resurrection Bay. As the aluminum hull slammed violently against the breaking 10-foot swells, she decided she was glad the boat was self-righting and tested in 30-foot waves. Its twin 500 horsepower Volvo engines roared over the noise of the storm as Tango and her assault team closed in on the beach not far from Dylan's cabin.

She had no idea how he and the ARCS team had escaped from the mine collapse. She shuddered thinking back to the terrible end-of-the-world groaning from the mine ceiling and rock falls as she and Boston tried to find their way out. This time the luck of the Alaska hunting guide had run out.

Her seven men were armed with the Sig MPXs for fighting in the woods, side arms and various other weapons according to the training regime of each soldier. But she figured it wouldn't be much of a fight, eight highly trained, well-equipped soldiers against a middle-aged, backwoods guide and two desk-bound scientists. If the order hadn't been to eliminate them all immediately, she could have taken care of the assignment in an afternoon.

Now she and her whole team were blasting across the bay in a million-dollar pilot/rescue boat during a millennial storm. Tango looked out of the spinning front windows and saw a wave break completely over the bow and flood the deck. She had no idea of how Sergeant Denver was going to get the boat up on shore far enough for them to disembark, but he assured her he could do it in a deep-water inlet near Lowell Point. It was the same steep pebbly beach where orcas regularly ran their bodies up to scratch their bellies.

Their assault plan was simple and direct. They would surround the subject's cabin, send in smoke grenades and kill the occupants as they left. Tango knew that simple plans were the best. However, she would not underestimate Dylan Baker. From her previous encounter with him, she knew he had a sixth sense for danger in his woods. By approaching their target from the beach, especially in this weather, he would be surprised. She wanted to avoid trying to kill him while he was loose in his own woods.

Boston unsteadily maneuvered to the seat next to her, "Ma'am, how are you enjoying this part of Alaska. It beats a coal mine, eh?"

Tango did not enjoy socializing with her team members, especially this team of fanatics. She would just as soon they not talk to her unless it was essential. However, she recognized that when soldiers depend on each other for physical and mental support, some socializing is necessary.

"I don't like this storm. What's with the weather around here?" Tango said, trying to sound like she cared.

"It's the climate change. Alaska has had more flooding and outlier storms than ever. The increased moisture in the air fuels the severity of the storms. It's pretty much the same all over the globe, but it's really visible here. Plus, cold water absorbs CO_2 much faster than warm, so the oceans around here are acidifying at a much higher pace than other places." Boston's voice rose as he spoke.

Tango knew he was about to go off on a rampage about the climate, so she redirected him. "I want you to check on Reno's pack. Make sure he has the 40mm grenades we talked about."

Boston nodded and made his way to the back of the boat.

"We're coming in. Hold on," yelled Sergeant Denver from the pilot's chair. The boat scraped against the pebbly beach and the bow rose out of the water. Two of the men leaped off the bow and onto the beach with lines to secure the boat. In the protected inlet, the surf was as mild as a lake, but the wind blasted rain horizontally and the wind tossed the treetops.

Tango figured that they would be able to surround Dylan's cabin and complete their mission within an hour. The Boss had told her to kill everyone and destroy the notebook with *the Scrubber*'s data.

She had divided herself and the seven soldiers into Team One and Team Two. She gave the order to move out, and her teams began their slog up the steep, muddy slope to their target. Over the storm, she could hear some crows make a fuss about their arrival.

104

BOLTON PUT DOWN his bourbon and scowled at his bill. He called a waiter over, "What's this charge?" he pointed to a small item on his restaurant tab.

"That's the table set up. It's expected in Argentina-style restaurants," the young, skinny waiter looked nervous.

"Well take it off. I planned my whole meal so the bill would end with the digits 22. It's my lucky number." Bolton's eyes bored into the waiter daring him to argue.

"I can't take it off, but I can adjust your bill to end with those digits," the waiter offered.

"Do it!" Bolton threw down his credit card, inwardly smiling at his private joke. The number "22" was symbolized by the ancients as "Good Fortune and Luck".

He pulled out his phone. It had been making a racket all evening, office security alerts probably due to Elizabeth staying late. He had enjoyed his huge steak dinner and didn't want to think about business. Now he had some dead time and punched in the unlock code to view email, stock reports, and his ever-annoying security alerts.

Bolton checked the log of card-scans to see who has been in the office. When Jane's card showed up, he became alert. *That bitch sneaked into the office without calling me! She was supposed to call me the minute she learned something.*

Something told Bolton that Jane did not just forget to call. She had entered her office long after the workday. She was up to something.

Bolton stood up so quickly, he knocked his chair backwards onto the floor of the restaurant. Without touching his espresso, he nearly ran out of the room.

JANE PUSHED against the solid closet doors. It was useless. They didn't budge a bit. She was trapped.

She sat down in the closet and noticed a tear splashing onto her computer screen as she debated which security code to use for the third and final attempt. *This is it. My job is over. I'll be charged with Bolton's crimes and not have the money to hire a great lawyer. Bolton was a soldier. He'd killed his enemies. Maybe he'll kill me. Make it look like suicide.* She started to shiver.

She shifted her weight knocking over the lock picks she had made with the little skewers and tools from the kitchen. One of the tools was a small pair of needle-nosed pliers. Looking around in the light from her computer screen, she noticed that the door pins were right in front of her. She could slide the door pins out of the hinges and push through the door.

Using the pliers, she grabbed the bottom door pin next to her knee and smoothly lifted it out of its track. *This is going to be easy.* Jane felt hopeful.

She jumped when she heard the phone on Elizabeth's desk ring. Jane started to move faster. She pulled on the middle pin but it seemed stuck. As the phone continued to ring, she gripped the pin tightly with the pliers and slammed her other hand against the pliers in an upward motion. No movement. It was like the door pin was welded tight. She wondered if she would get out.

The phone rang again for the fifth time and the answering machine picked up. After the greeting message Jane heard a foreign-sounding voice. *Colonel Bolton? This is Roule from the Asado Argentina Restaurant. You left your credit card here, and you didn't sign for the bill. I put your lucky number as the last two digits of the bill as you asked. If we don't hear from you in 24 hours, we'll put through the charges for the meal plus a 22% service charge. Twenty-two is my lucky number, also.*

Jane stopped working on the door pins and picked up her laptop. After entering 22 as the security password, Jane winced when nothing happened. *Did it work?* Suddenly, a list of files scrolled across her screen. All of them were put on the server within the last 12 hours. It looked like Bolton was purging. Jane pulled the icons for all the files onto her laptop and watched a scroll bar slowly start to move. The files were being copied to her laptop.

As the scroll bar made its slow journey to the right, Jane went back to work on the second pin. It wouldn't move. She tried to use a hanger to leverage the pliers upwards, but no movement. She looked down at the hinge that was so easy to remove, and noticed that the door had shifted so the parts of the hinge were no longer lined up.

She pushed on the door so the bottom hinge parts lined up and the pin from the middle hinge slipped easily out. The door fell crooked with just one pin, the top one, holding it on.

Out in the hall, the elevator made a *ding* and Jane could hear Bolton's purposeful strides in the hallway followed by a *beep* telling her that he had slid his card through the card reader.

105

TANGO'S TEAMS SIGNALED to her that they were in position and ready, Team One consisted of herself and five others in front of the cabin. Team Two was made up of two men, Memphis and Salem, with SIG MPXs covering the back.

Tango toggled her radio, "Phoenix, did you disable their truck?"

"Roger on that, slashed tires."

"Let's get started," Tango said.

She gave a nod to Reno. He lifted up his M320 grenade launcher and fired the first of two smoke rounds. When he pulled the trigger, the gun fired with a loud *tick* sound, not a boom like a shotgun. The explosion came when the first round crashed through the front window.

Reno loaded another round into the single-round tube fastened to the rail under his assault rifle. Another *tick* and the second hit the window frame and bounced into a flower garden in front of the cabin.

Moments later, smoke billowed out of the front window and from the garden. From inside the cabin, a hand came up and tossed the smoking grenade out the broken window.

Tango signaled and Phoenix brought up a megaphone, "This is the Alaska State Police. Come out with your hands up."

The team had only two smoke rounds for the M320. No one came out of the cabin, but someone inside had thrown the smoke grenade out the window. She knew they were in there.

Phoenix again brought the megaphone up, "You have 10 seconds to comply. Come out now and no one will be hurt."

Tango counted to 10, and then gave the signal to open fire. Six sub machine guns opened fire nearly at once. Since all the weapons were equipped with the silencers, nearly no sounds other than the rainstorm could be heard through the forest.

Wood chips flew off the cabin as if the logs were trying to rid themselves of wood and the log home's windows soon disintegrated under the hail of hundreds of bullets.

Tango ordered her team to hold their fire. She could see no movement in the house. She hated the way the logs absorbed the bullets. She doubted if any rounds made it through the thick walls.

The air smelled of gunfire, rain and mud. The rain slashed down and the trees tossed as the storm pounded the woods. She ordered the grenade launcher loaded with an M430 armor-piercing cartridge, which can penetrate 51 mm of steel plating.

The Team One watched Reno aim at the door and heard the *tick* followed quickly by an explosive flash and boom. A hole the size of a grapefruit opened above the door of the cabin and burst into flames.

Reno gave Tango a questioning look, "Tango, is that house built out of reinforced concrete? That rocket should have knocked a huge hole in it."

"It's made of 20 inch Alaska spruce. The soft wood absorbs kinetic energy. It will take more than one M430 to dislodge the residents."

Phoenix again told the occupants to come out with their hands up. Tango waited for the count of 10, and then signaled for another M430 round to be fired. After the *tick*, a flash, an explosion and a new hole appeared on the other side of the door.

Tango knew Reno was trying to hit the door. At 100-yards, the storm threw off the accuracy of the rocket. From inside the house, a container of water splashed on the first flaming hole and extinguished most of the fire.

Tango ordered the others to give cover fire as she and Reno moved forward. This would be repeated until the front-door team was within 50 yards of the cabin.

She would then order the grenade launcher loaded with incendiary rounds. These rockets were banned by many militaries because they are considered to be a chemical weapon. A drop of white phosphorus on a person can cause third-degree burns, the fumes are lethal and the toxic smoke it produces can cause a painful death. After one of those exploded, anyone left living would be down on the floor writhing in pain.

Tango thumbed her radio, "Team Two, we are rushing the house in 20 seconds. Be ready. If they are alive, they may run out the back. Shoot to kill."

106

JANE NEARLY CRIED out when the door crashed open and Bolton strode in bringing a nearly palpable amount of energy into the quiet office suite. She peered through the ventilation grate into the outer office. Her face on the dusty closet floor carpet, she could see him from his waist down. His body conveyed anger and purpose. She watched as he left the outer area and charged into her office. Soon he seemed to be searching the entire suite.

When he exited his own office, she could see he carried the long-barreled antique revolver as he crossed the room to search the kitchen area.

Jane held her breath and hoped the crooked closet door would not attract his attention. After a loud clattering in the kitchen where Bolton was evidently searching the storage cupboards, he strode into the front office and paused. Jane thought he might sense her presence or hear her hammering heart.

Jane suddenly wondered if the light from her laptop screen could be seen from outside the closet. She thought about closing the laptop, but wondered if that would stop the download or make an audible sound to draw Bolton's attention. She looked at the screen and noticed the download was nearly finished. *Will this make a sound when the files are fully downloaded?*

In a near panic, Jane tried to remember how to turn down the volume of the computer. Bolton left the room and reentered her office. She could hear drawers slamming as he searched her office again. Her laptop made a loud and cheerful *ping* as the screen displayed the message, *Download Complete.*

At the sound of the ping, Jane could hear Bolton stop his search. When he entered the reception area, he walked in stealth mode with the gun held ready. She held her breath as he stood in front of the closet. She nearly wet her pants when he fumbled with the padlock. *"I hope he doesn't notice the door is canted!"* worried Jane.

After what seemed like minutes, he returned to his office and made a call. She could hear him punching in a number on his desk phone. She wondered if this would be the time to pull the last pin from the door and rush out. She could hear his side of the conversation due to his office door being ajar and his booming army voice.

"Krish. We got a problem. I think the woman who runs Yeung's Super PAC is going to the authorities to report our activities. She's been making recordings and gathering evidence."

After a pause Bolton's urgency turned to anger, "It's your problem too. She videoed most of that meeting in my penthouse and god knows what else."

Bolton listened then spoke without anger, but with a voice that conveyed he expected to be obeyed. "Listen,

Krish. I don't know if she has any proof in her possession. I erased her recordings, but we need to stop her. Who do you have who does wet work? Get over here as soon as possible. I don't want to talk about this on the phone."

Jane's heart rate shot up at the words, *wet work.* Looking at her computer, she started typing. She had to upload the video to a safe place so Bolton could not seize her computer and take away her proof. She could think of only one website to put the video, the Super Pac's website. After entering the login codes, Jane started the upload.

A loud *bang* made her jump and nearly cry out. It sounded as if someone had just kicked the door. "Jane, I just looked over the security footage from the last hour. You are locked in the closet," Bolton said. "Just where I want you. Now slide your phone under the door."

107

THE BUTT OF DYLAN's 7-pound Marlin hit Memphis in the back of his head, just at the center of his helmet. Even in the storm, Dylan could hear the *thud.* When the unconscious soldier pitched forward, Salem stood up, his jack-o-lantern teeth yellow in the firelight, and stared at his companion who was slowly writhing face down in the mud. His look of puzzlement changed to surprise when he saw Dylan standing in front of him, rifle pointed. Salem brought up his Sig.

Dylan fired his rifle at point blank range into Salem's chest. The .458-inch bullet traveling at 631 meters per second destroyed the man's armor and threw him back against a tree. Both men guarding the back of the cabin were down.

TANGO GAVE the order for the frontal attack to begin. As her soldiers initiated their assault procedure, she toggled her radio again, "Team Two, do you read me? We are launching incendiary grenades. Be ready for a run out the back."

No answer came. Those guys had probably turned down their radios so they could discuss climate change. She hated working with fanatics, but this was what The Boss had provided her. Tango gave it one more try, "Team Two, respond immediately." She looked up at Reno and saw him put a red laser dot on the door and pull the trigger.

Following the *tick* came another explosion at the front of the cabin. This time flames covered an area on the corner of the house the size of a military jeep. The grenade-launcher fired three more incendiary rounds in close succession. One of the rounds went through the front window and the interior of the cabin burst into flames. A high-pitched scream followed the flames as they burst out of the windows.

Tango's soldiers moved forward under cover fire as the front of the cabin burned with apocalyptic fury. The wind-driven rain did not dampen the flames' passion to burn as they swirled and grew like an evil red genie surrounding the cabin and attempting to lift it into the gigantic Alaska storm.

Once her soldiers were in position, she and Concord threw hand grenades through the open window and took cover. The blasts blew out all the remaining windows and launched a flaming Hawaiian canoe out an upper window and off the cliff face toward Resurrection Bay leaving a wake of sparks and smoke.

Tango toggled her radio, "Team Two, report now."

USING HIS KEY, Dylan opened the damaged back door and called in, "Sully! Taylor! Don't shoot! It's me. Are you alright?"

From the smoky kitchen Taylor called back, "You're back! Sully's hurt. Something burned his leg. We crawled back here when he got burned."

Scrambling into the smoky room, the devastation in his home shocked Dylan. Flames and a bitter, acrid smoke boiled into the kitchen from the living room. In the flickering light of the fires, he could see a golf ball-sized chunk of flesh gone from Sully's upper thigh. The area around the wound was charred as if a cutting torch had bored into him. Sully cried as he hugged a two-liter pop bottle, "It burns! It burns!"

"You guys need to leave here now!" he looked into their sooty, terrified faces. "We just barely have enough time to run before the soldiers out front get us," Dylan yelled over the howling of the fire and storm. "Can either of you ride a motorcycle?"

"Why are the State Police shooting at us? Are these more of the fake state police?" asked Taylor.

"It's not the police, but a bunch of killers. We got to get out of here," said Dylan.

"Why not take the truck?" asked Taylor.

"The tires are slashed," said Dylan.

Sully looked up and sobbed, "I can ride an electric bicycle, but not a motorcycle."

"Good enough, then you are taking my Rokon up the hill and make a 911 call for help. Do you have a phone?" Dylan asked.

Sully nodded and held up his phone.

"Taylor, you come with me. We'll hide near High Rocks until help comes."

"What's a Rokon? Is it like a Harley?" Sully asked as a sagging overhead beam dropped a shower of sparks and cinders on them.

"We got to get out NOW! Grab those rain slickers on your way out!" Dylan yelled. "The motorcycle is in the shed."

Crawling, Taylor and Sully followed Dylan out of the flaming cabin and into the dark and raging storm.

OUT FRONT, Tango shouted over the storm and ordered Concord and Denver to go around back and check on Team Two.

108

LEAVING THE CABIN, they crept past two soldiers—one obviously dead. Sully squealed, "He's dead! His chest is all caved in!"

"Sully, be quiet," Taylor urged, also alarmed by the dead man. The other soldier moaned as they continued past him and into the woods.

Not far from where Suzie's pickup listed to the left on flat tires, a large metal utility shed reflected the flashes of light from the burning cabin. Dylan unlocked the doors. Just as he was pulling them open, a huge shiny black creature came charging down out of the blackness of the night.

Sully screamed when the moose thundered right past the shed. "Sully, stop that!" Taylor hissed. "You'll give away our position!"

"But that big black thing ran right at me," wailed Sully.

"I've never in my life seen a moose run in panic like that, especially at night. They don't run at night," said Dylan. "Something really scared it, or pissed it off."

Once inside the shed Dylan pushed the doors wide open so flashes from the burning cabin could light the area. Dylan pulled a tarp off a large form in the back revealing the oddest motorcycle Taylor had ever seen. It had wide tires that looked like they came off a tractor and an engine that seemed like it was stolen from a lawnmower. "What's this?" she asked as she wrapped Sully's burn using a roll of gauze from a first aid kit.

"It's my Rokon; a two-wheel drive motorcycle that floats over the forest floor. It doesn't tear up the forest like a dirt bike. It goes pretty slow, but it can pull a cart or sled so I can help a client get a moose or bear carcass out of the forest." Dylan showed Sully how to use the simple machine and told him to head east and uphill so he could catch a cell phone signal.

Dylan explained to Sully where he and Taylor would be hiding, but told him to head into town after he made the call. His leg needed medical attention.

"What's this? An antenna of some kind?" Sully pointed to a set of Y-clamps mounted on the handlebars.

"That's a gun mount," explained Dylan. "It allows a hunter to have easy access to his rifle."

Sully paled, "I got to go back. I left my shotgun in the kitchen!"

"No. It's too dangerous to go back," said Dylan.

"But I need a gun!" Sully sounded close to breaking down.

Dylan put his rifle into the clamps and secured it. He spoke calmly hoping Sully would relax and focus. "This is better, you don't need to pump it like the shotgun, just bring down the lever to get a shell in the chamber. Like a cowboy. You have three shots left and the first shot is already in the chamber." Dylan loaded another cartridge into the magazine. "There. Now you have four shots."

Taylor put a hand on Dylan's arm, "You need a gun. Want my pistol?"

Dylan answered her by taking a wicked-looking compound bow off a wall rack. He opened a drawer and pulled out a quiver of arrows, the broad heads reflecting light from their razor-sharp edges. "This is all I need." Inwardly, Dylan wanted his Marlin, but he also saw a need to settle Sully's panic. The bow would have to do.

Dylan passed out headlamps to everyone, "Only use the red light. It will preserve your night vision and be much harder to spot by those soldiers."

Dylan tossed a coil of rope and a first aid kit into a daypack, and then he and Taylor pushed the bike out of the shed and deeper into the woods. While they were doing this, they thought they heard another moose or other large animal running through the woods toward the bay.

"Sully, once you start this, the lights will come on. You'll need them, but the killers might see them, too. So don't stop. They can't catch you if you keep moving," said Dylan. He hoped he was right.

Soon, Sully started the bike and headed into the inky-black forest at a walking pace—the Rokon's top speed.

Dylan took Taylor's arm, "Those guys will follow him. He's noisy and has that light, but they can't catch him, unless he stops."

Dylan led her uphill into the darkness leaving the flaming cabin behind. Dylan peered into the darkness. Somewhere, up there, a force was terrifying the animals and causing them to flee downhill toward the bay. He knew that whatever was frightening the game, was waiting for them up there.

He tried to think of what could be scaring the animals. Dylan had never seen a forest full of panicked animals save for a big Oregon forest fire he'd got caught up in years ago. There was no fire up the slope. In a way, the rain was helping them to hide. Dylan knew that the night-vision gear effectiveness was greatly reduced in the rain.

The rain was so intense, Dylan felt as if someone was above him pouring buckets of water over him. He'd never experienced such a fierce rainfall before. It was an eerily warm rain for spring. For a moment, he wondered what all this warm rain was doing to the snow cover on the Harding Icefield just a few thousand feet higher in elevation than his present location.

Dylan knew that the icefield was thousands of feet thick and about a third the size of Rhode Island. In the summer, Dylan led treks to and on the icefield just a mile from his cabin. It stretched out farther than an eye could see, and spawned massive but shrinking glaciers.

WHAT DYLAN DIDN'T KNOW was that throughout the past decade, a gigantic Manhattan-sized sub glacial lake, had formed just a few hundred meters under the frozen surface of the ice sheet. As the warm, spring rain hit the icefield and rapidly melted the snow covering, the waters gushed into the lake with maniacal force, straining the ice dams that held it back.

Anyone unlucky enough to be standing on the icefield would have heard a earthquake-like roaring and an eerie moaning as the ice dams began to weaken and the entire ice shelf shuddered.

GOVERNOR LEOLA Bates-Hardy pulled her attention back from the Anchorage cityscape as the flashing lights of the emergency vehicles captured her peripheral vision. She increased the TV volume to hear the reporter's update on an explosion at the Botanical Gardens. *I hope terrorism has not come to our sweet town!*

What had been scheduled as a quiet time to catch up on correspondence for the governor, had become a powerfully disquieting evening. First, Harold Cheek was a "no show" for several important meetings. And now a report of an explosion and an unidentified body at the Botanical Gardens.

"Mr. Cheek's phone isn't turned on. He signed out of the office and his parking spot is empty." The receptionist shuffled through several computer screens. "I have a calendar entry here that he did speak with Anja Hart today. Perhaps they scheduled a meeting."

"Anja Hart?" The governor appeared perplexed. It would be a 90-degree summer day in Anchorage before you would find those two relaxing in a local pub! "Have you tried to contact Ms. Hart?"

"She's not answering either. Her phone goes straight to voicemail."

This wasn't making any sense! Leola walked over to the cabinet and opened a Dr. Pepper.

109

CONCORD RUSHED UP to Tango. "Those bastards killed Salem. He's shot in the chest. Dead."

"What's the situation? Are the targets dead? Is the notebook destroyed?"

"I don't know. Some or all may have escaped. The fire is too hot for us to check the cabin for bodies. Salem was shot at close range so I think at least one got away."

"Anything else?" Tango asked.

"We found Memphis laying next to Salem. He's drifting in and out of consciousness due to a head injury. He's in no condition to fight."

"What the hell?" Tango said as several deer ran downhill past the blazing cabin towards the bay. She'd never seen game run toward a fire before. Quickly she refocused.

"Make Memphis as comfortable as you can. You and Boston are in charge of making sure no one can get to town. Set up a block on the path to town. Be ready to join us if we corner them. The rest of us are going to search the woods around here for anyone hiding." Tango gestured to her men, and they initiated a search pattern.

"Look!" Reno pointed to lights bouncing in the woods.

"That's not a man running. It's a vehicle. I can see a red taillight. It's going very slow." Tango looked at

Reno, his long red-gold hair flowing from under his helmet looked black in the rain. "Can you catch him?"

Reno smiled, "No problem." He tossed his Sig to Concord, pulled out his sidearm and started jogging after the light. He figured he'd catch the escapee in less than ten minutes.

Tango addressed her remaining troops, "Resume your search pattern. Keep checking any high ground." Tango wanted to get Dylan. Now he was more than a target to her; he was a trophy.

THE VAST SUB GLACIAL LAKE, lying hundreds of meters below the surface of the Harding Icefield, continued to rapidly fill from melt water and rain runoff. With no surface drainage path, huge shallow ponds on the surface of the icefield opened up subsurface water channels directly filling the growing lake. In a normal season, these ponds would freeze in the spring and slowly release water through the short summer. This was not a normal season.

As the Harding Icefield sub glacial lake expanded, the pressure at the bottom became powerful enough to lower the freezing point of the water below the temperature of the ice forming the dam. Liquid water began flowing through cracks in the ice dam. This flow caused friction that generated enough heat to expand the cracks. The expansion allowed more water and even greater friction. The ice dam slowly began to change from brittle ice to slushy, frozen plasma. Plasma can't hold back water.

This process, called *glacial lake outburst* or jökulhlaup, would soon have water bursting through in a catastrophic event. A jökulhlaup can send a 1,000-foot wall of water at speeds over 800 miles per hour down a slope.

DYLAN AND TAYLOR moved uphill toward the icefield, every step in several inches of water. A heavy blanket of water covered the entire forest floor. Dylan noticed Tonsina Creek, which he had crossed on stepping stones a few hours ago, was now a furious, tumbling maelstrom rushing madly down a streambed carrying all manner of debris, trees, bushes, and a drowned bear swirled past. They changed their route to High Rocks to avoid the river.

Dylan suddenly thought of Sully; he needed to cross Spruce Creek to get to Seward. It would be impossible. If Tonsina Creek was this bad, Spruce Creek, located in a sharp cut between two mountains, would be far worse. Spruce Creek's outlet was near the little community of Lowell Point. Surely the Seward Police Department was busy evacuating everyone. Dylan's group would get no help.

110

BOLTON PICKED UP JANE's phone from the floor and turned it off. "Come out of that closet," Bolton ordered. His voice sounded like he was Moses telling the seas to part.

"There's a padlock on the door," Jane called out. She worked quickly to drop the pins back into the hinges.

"How did you get in there then?" asked Bolton sounding curious

"It was unlocked. I entered. Elizabeth locked it not knowing I was in here," Jane said, then started to describe in detail how she got locked in the closet to keep Bolton busy while she checked her laptop to see if the video of Bolton's meeting with his oil partners was uploaded onto the Super Pac's website. This video would show the world that Bolton, and not Jane was the criminal.

"Why did you go into the closet in the first place? Are you pulling files from our office server?" Bolton asked.

"It seems a sneaky thing to do. Are you working against me now?"

"Working against you? Why would I do that? I need your recommendations and contacts to get my next campaign job," said Jane.

"If you are taking files off the server, you will need to pay for them. I know just how you can pay me," Bolton's voice had taken a creepy tone.

JANE BEGAN explaining to Bolton, what sounded like a really lame reason for her to be in the closet. As she talked to Bolton, she noticed that the video and dozens of public comments were waiting for Bolton to approve them before they became visible to the public on the Super Pac's website. Her heart fell when she realized he would be very unlikely to approve a video he had not viewed.

"I guess we'll need to wait until Elizabeth comes in tomorrow to get me out of here," said Jane. "You can come back in the morning and watch her unlock the door."

"Why would I do that? The key is right here in Elizabeth's paperclip bowl." As the key and lock rattled outside the door, Jane quickly put her laptop in the back of the closet behind the server and hid the lock picks in her pocket. She hoped Bolton would not search the closet and find the laptop nor pat her down and find the lock picks. She cringed at the thought that he'd pat her down.

When Bolton ripped open the door, Jane found herself making a squeaking sound.

Bolton looked down at Jane sitting on the closet floor with dust on one side of her face. She looked like a child playing with a pair of pliers.

"Now I know just what to do with you," Bolton smiled showing his absurdly white teeth.

111

DYLAN AND TAYLOR stood at the base of High Rocks, the cliff face where he had taken a call from Bolton just weeks back. It seemed like years ago, but there had been cell reception then, and there might be some now. The cliff also offered some shelter from the wind, but the rain relentlessly poured over them.

"Taylor, I don't think Sully is going to get us any help. Even if he gets a call through, I'm sure the police are busy with all kinds of storm-caused emergencies. The whole town of Seward is surrounded by creeks that are undoubtedly overflowing," Dylan found himself shouting.

"What should we do then? We're not soldiers. I have a pistol, and you don't even have a gun. How can we fight a small army?" said Taylor.

"I can fight. These are my woods, and I only saw four other soldiers," said Dylan.

"I can fight, too," said Taylor. "Give me your phone. If these guys are trying to suppress the truth about *the Scrubber*, we have pictures of Sully's report on your phone. I can email or text this to someone, if I can get a better signal."

"Do that. This rain may be weakening the phone signal or with everyone in town calling someone, the network could be overloaded. Keep trying."

"I'll keep trying, but your phone is nearly dead. You haven't charged it since we were at the mining camp." Taylor put the phone under her poncho. "Can we go higher? Maybe there's a better signal higher."

"We sure can. I'm going to check on those guys who are following us. You move around to the right. These rocks are less steep there and you can scramble up. You can use the light from the burning cabin to see. I need your headlamp." Dylan slipped her headlamp off and disappeared into the storm.

TANGO HAD turned up her radio so she could hear her troops over the rain. A call came from Denver, "Tango, I see something. It looks like a red headlamp over by some tall rocks."

"What's your position?" Tango radioed back. She knew there were big rocks to the west, and Dylan loved big rocks.

"I'm on the far west side of our search pattern about 800 yards uphill from the cabin."

"Everyone converge on Denver's position. We'll decide on what kind of attack to execute. Remember, these targets are armed." Tango's heart rate accelerated. She would now get Dylan. He would not be the one to make her mission a failure.

As she moved through the rain, she marveled at the sheer amount of water falling from the sky and flowing over the ground. Her wet feet sloshed around in her "leak-proof" boots. Her other gear seemed to be failing, too.

She was disgusted over the performance of their night-vision goggles. She remembered in Afghanistan, when a helicopter landed and threw up a cloud of dust, the NV goggles didn't work that well. In this rain, nothing but her trusty Sig seemed to work. It pleased Tango that the burning cabin provided just enough light for her team to complete their mission. She looked back at the cabin and decided the flames were getting smaller. She needed to hurry.

She'd shot a porcupine a few minutes ago. It was one of dozens of animals moving toward the bay. It gave her the creeps. She wanted to eliminate her targets and get back to the boat as soon as possible. She hated Alaska.

"Tango, I have a shot," Denver radioed back. "Permission to engage? Concord is with me, and we can both open fire at once." Tango wanted to be the one who got Dylan, but she knew the mission's success was more important than her own desires.

"What do you have?" Tango radioed back.

"I can see them climbing a big tree. A red headlamp is giving them away."

"Denver, you and Concord have permission to engage. Be aware that I am coming in from the east." Tango figured she'd be in the boat soon, but she had called in Phoenix and Boston. No need to guard the road to Seward.

"Keep your channel open." Tango wanted the others to hear the battle.

"Phoenix to Tango," he sounded stressed.

"Go ahead, Phoenix."

"We stopped an ATV here and one of the occupants claims to be The Boss."

"Give him your radio. I can confirm the identity," said Tango, surprised that anyone would voluntarily come up to this wretched site this time of night in the middle of a rainstorm.

After a brief conversation, Tango was certain The Boss was now on site. Hearing The Boss' unmodified voice surprised her, but the voice knew all the code words. It really irritated Tango, because a new order came with The Boss: *Don't kill Dylan.*

Using hand signals, Denver and Concord closed the distance to the bouncing red light on the huge Sitka spruce. Through the rain, Concord could barely see Denver's hand countdown the order to open fire in 5, 4, 3, 2 . . .

Just then Denver fell to his knees and grabbed his throat. In the dim, flickering light, it looked like a stick was protruding from Denver's neck. He fell backward into the mud. Moving forward, Concord could see an arrow sticking out of Denver's throat. *An arrow?*

Concord sprayed the bobbing red light with automatic fire, but the archer wasn't near the tree where the lights were fastened.

Dylan was a mud-covered Alaskan holding a Bear Motive 6 bow with a 30-inch draw and 350 feet-per-second speed. The razor-sharp broad head passed completely through Concord's neck severing his carotid arteries and adding his blood to the oversaturated earth.

Dylan scavenged the radio and submachine gun off Denver's body then retrieved his light from where it dangled in the nearby tree. No one fucked with Dylan in his own woods.

112

TAYLOR CLIMBED THE ROCKS blindly feeling her way along as the sky imperceptibly brightened in a dim grey, rainy dawn. Far below, the burning cabin cast an eerie feeble flickering light on her climb. She felt as if she were scrambling up a shallow waterfall as she ascended the rain-washed stones.

Each time she thought she was able to get another ten feet in elevation, she ducked into her poncho and checked the phone. It showed one bar that seemed to fade in and out. The screen fogged with condensation from inside the phone, and she wondered how long it would continue to function.

She told herself that when she got two bars, she would attempt to send the message again. She hoped there was enough power left in it. The phone had shut itself off once already and was blinking a red warning: *Charge Battery*.

Nearby a huge tree slowly started to lean downhill, then collapsed into the rain-drenched darkness. Turning her back to the rocks and wind, Taylor ducked into her poncho and checked her phone again. Two bars. She hit the send key and watched a progress bar start to move across the screen of Dylan's phone.

The phone went black just as the bar reached the end. Taylor didn't know if the text made it out. The phone would not turn on. It was dead.

BELOW THE ROCKS, Dylan held a Sig MPX with a fresh magazine and listened to the radio headset. He heard the people who burned his home and tried to kill him and his friends. "Denver, this is Tango, status report?" Pause. "Boston, status report?"

Dylan toggled the mic, "Tango, this is Dylan. It's over. You can go now. No one else needs to get hurt."

Radio silence.

"All units converge to previously stated location. Proceed with utmost caution," came Tango's reply.

Dylan crouched in some bushes, invisible to NV goggles or flashlights, "Tango, your men are dead. If you and the rest keep coming, they'll be dead. It's over. This is my mountain, and it's time for you to leave."

Tango radioed. "If you surrender now, you can live."

"Tango, your remaining soldiers cannot see me. But I know you came here because we learned *the Scrubber* is a fraud. We just texted a full report about this." Dylan did not ask Taylor to whom she was sending the text. "It's not a secret any more. There's nothing for you here. Leave now."

Radio silence.

"Dylan, where did you send the text? Who has it?" Tango asked.

Deep in his bushes, Dylan thought he heard someone approaching from behind, but with all the water noise, it would be impossible to know for sure. What he did know was that soldiers walking with NV goggles would be nearly blind. The goggles were confused by a little rain, and this was a lot of rain. Besides he was in deep cover. For a moment, Dylan felt invisible and safe.

"The secret is out. You have no reason to stay," *Reason to stay,* the words triggered unbidden images of his wife, Suzie. He should have given Suzie more reasons to stay. *It's my fault she left me.* An unavoidable sob rose from somewhere deep in his chest.

A loud splash behind him took him by surprise. The cold gun barrel of a Colt M4 carbine pushed against his neck instantly stopped his self-recriminations.

"Drop the Sig," said a familiar male voice.

IN AN ANCHORAGE OFFICE, a cell phone vibrated: *Incoming Text From Dylan Baker.*

113

THE HALF-FULL COFFEE cup had been replaced by a glass of wine. Governor Leola Bates-Hardy needed some time to process the latest information. She leaned back in the chair; her mind spinning with the implications.

Leola pulled out a notepad to make a list; a technique that often helped her see the facts. First she jotted down the known facts. A yet to be identified, badly burned, adult male discovered at the Botanical Gardens killed by an explosion.

Word from the police department indicated a badly burned and unreadable drivers license. Also, Harold Cheek was missing. Cheek's calendar had indicated a call or meeting with Professor Dave White and a phone meeting with Anja. No location listed. This was odd for the meticulous Harold Cheek.

What if Harold Cheek was dead? That thought felt like a punch to the stomach. Cheek was a brilliant strategist responsible for saving the governor's political life several times. Although a little on the edge at times, Cheek had earned Leola's respect.

The governor returned to her notepad. Next was the list of questions. Why would Dave White schedule a meeting with Cheek? Dave's admission during this morning's meeting that much of the contradictory scrubber performance data had been omitted from the final report was a devastating blow to the governor's campaign strategy.

Future research grants were strongly linked to the reduction of CO_2 in coal plant smokestack emissions. Did Dave schedule a meeting with Cheek after our meeting? Could something have happened to Dave? Why wasn't Dave returning her calls? The Alaska State Troopers were trying to track down Cheek and White, but so far, no information.

Leola set aside the notepad and returned to her glass of wine. Dave White! He had put her in a terrible position. Integrity and transparency, she had strongly reinforced her key principles and her dismay with Professor White during this morning's meeting. Now she would need to consider the implications to her campaign. *And where was Cheek?*

And where was Bolton in all of this? Obviously, discrediting CO_2 emission reduction technology was a step along Bolton's path to success, as would be the disappearance of his nemesis, Harold Cheek.

No more negatives! Leola would not compromise her foundational principles. She would win this election on the issues and not by tarnishing the people in Greg Yeung's campaign.

Shirley Ahn grabbed Leola's attention. "As you requested, I attempted to contact Dr. White again. I was unable to make contact and inquire about Harold Cheek. I left another message on his voice mail to call this office ASAP."

"Dr. White is mysteriously gone, too?" *What the hell is happening?*

114

"PUT YOUR HANDS behind your neck. Interlace your fingers," said the voice.

Dylan did as he was told. *Where have I heard that voice before?* Then it came back to him; it was Roger,

Anja's bodyguard. *What the hell is he doing up here?*
Dylan wondered if Roger was working for someone
besides Anja. Maybe he wasn't guarding her after all.
Maybe he was actually keeping an eye on her.

"Roger, it's me, Dylan. Anja's dad."

"Put the bow down," Roger's voice conveyed no
surprise or warmth.

"How did you find me in all this darkness? I'm
covered in black mud in a black forest," Dylan asked
turning to face his captor. What he saw was a tall, fit
man Roger's size, but wearing a helmet that looked like
it came from a science fiction movie. Even as he spoke
a part of his mind noticed the rain and wind were
abating. The storm was easing.

Roger tapped his helmet with its goggles set in the
down position, "Stage 6 night vision. Even the army
doesn't have it yet. I can see through rain. Now walk
over to that rock. Put your hands on it and spread your
legs." Roger did a pat down search and found no other
weapons.

Suddenly, Tango came up from behind, "You asshole.
You could have given us those NV helmets."

Roger looked at her, "You didn't ask for them. The
Boss gave you everything you asked for."

"I have four casualties that could have been avoided.
These were troops you recruited and assigned to me.
This loss is your fault," Tango yelled as she turned on a
powerful headlamp. The brightness caused Roger to
flinch and remove his helmet.

"Where's Anja?" Dylan asked.

"Shut the fuck up," said Roger as he turned on his
headlamp.

"I'm here. I'm The Boss," Anja stepped out of the
shadows followed by Andrew, who also carried a Colt,
M4 carbine.

"Don't get too close, Ma'am. I searched him, but he's
dangerous," said Roger protectively.

"Dylan won't hurt me. Would you, *Dad*?" Anja said the last word dripping with sarcasm.

Dylan felt confused and weak kneed. *Is Anja here to help us? Why are Anja and Roger with Tango? Why would they want to keep the failure of the Scrubber a secret? Did they know all along it was phony?*

"You look confused, Dylan," said Anja. "But you needn't be. We must keep the Mountain Club strong. To do that, we need public support and money."

"I don't understand," said Dylan raindrops glistening on his mud-spattered poncho.

"Dylan, you screwed up our plans. We were on track to making the Mountain Club relevant. It was about to step up beyond just being an advocacy organization. We were going to force change on the public, and they were going to welcome it."

Dylan felt suddenly dizzy. The world was not right. He tried to focus on Anja's words.

"We must save Mother Earth. With milk-toast politicians like Leola Bates-Hardy, environmental change would happen at a glacial pace. Too slow to save our Mother. If Bates-Hardy, and people like her were elected, the public would think something significant was being done about climate change. But it would be too little and too late.

Something was tugging earnestly on Dylan's attention. Calling him. Shouting at him, but he had to push it away to listen to Anja, his daughter. If he listened carefully, maybe he could save her. He spoke gently, "If Yeung gets elected, how will that help things?"

Anja spoke as if she were addressing a child, "If the governor put all her eggs into *the Scrubber* basket, the public would be shocked that either the governor was too gullible or too dishonest to enlighten the voters. Once the truth about *the Scrubber*'s ineffectiveness was released, then Greg Yeung would be elected, and he'd cause more damage to the environment with all his deregulation plans.

"Yeung would boost development of natural gas and other petroleum products. The public would rise up in anger. Yeung's recall would be one oil spill away, and oil will spill.

"Donations and volunteers would pour into the Mountain Club. Yeung could be quickly recalled and a powerful environmental candidate would be elected. The new governor would be in a position to undo much of the damage of the past 50 years. She would have a bully pulpit to convince Alaskans and others of the urgency of doing something."

The background noise trying to get Dylan's attention was nearly screaming at him. It took all his will power to ignore it as he spoke, "But Anja, why are you here? Why cause harm to me and my friends?"

"When I passed along that notebook with *the Scrubber* specs, I had no idea that a simple backwoods guide, would be able to find information in it that would prove it wouldn't work. Christ, Dylan you are not that smart. Still, you surprised me with how far you and your friends have come. Tango said you are dangerous, but you look pitiful standing there, covered in mud against the rock wall."

"But Anja. You are willing to kill people over this election?" Dylan felt he should cover his ears to avoid the incessant alarm sounding in his head.

"It is unfortunate that some people had to die to get this plan moving, but what are a few deaths in comparison to saving Earth? I hated ordering the deaths of Professor Carlyle, Harold Cheek, Zhang, Ethan, and now Sully, Taylor and you. However, in a war, the side of righteousness must expect some collateral damage. What battle could be more important than saving the earth? "

DYLAN LOOKED at Anja, his tears invisible on his wet face. *She's crazy.*

"So we need to know where you sent the report on *the Scrubber*. We have people in places to intercept it. It needs to be released on our schedule," Anja said, her voice calm and logical.

"What of Sully and Taylor? They are out there somewhere," said Dylan. "Call off your eco-terrorists."

"Dylan. Didn't you hear what I said about collateral damage? I'm sad. I truly am, but Earth is more important than any of us. You are too simple a person to understand a grand plan to save Earth. I know you mean well with your 'morals' and all, but this situation is far beyond your abilities to comprehend," explained Anja calmly.

"If I tell you where I texted the report, will you promise to not harm Sully and Taylor?" Dylan negotiated. He knew Anja did not feel constrained by promises, but he had to try.

"You are bargaining with me?" Anja smiled. "Well of course then. Tell us where you sent the message, and we'll not harm anyone else." Her tone conveyed poorly concealed sarcasm.

Dylan stood feeling alone in the darkness, a large stone wall behind him, water now just dripping off the trees but still flowing over the ground and around his feet. In the night sky, rain clouds gave way to mist with a silvery, nearly full moon lighting the sodden woods. Then he noticed the water flowing over his boots was a milky green.

This was glacial melt water! The voice in his head erupted, *The glaciers are melting!*

115

"TAKE MY HAND, little girl. I know something fun we can do together," Bolton transferred the revolver to his left hand and held out his right.

Jane slapped his hand away, "Knock it off. I found out something that guarantees Yeung's election. I'm not working against you." She stood as tall as possible and feigned confidence as she strode out of the closet past Bolton, dust covering one side of her face.

Bolton hurried to catch up and tried to put his arm around her small shoulders to draw her to him, "I don't really care. I want to play a touching game with you."

Jane shoved his arm away and headed for his office. "Quit clowning around, I need to show you something. The governor is basing her campaign on two things, global climate changes effects on Alaska, and the carbon dioxide mitigation technology developed by Dave White."

I need to distract him, Jane thought. Once in Bolton's office, she stood before his open laptop. The security screen glowed as it waited for the unlock code to be entered. "Open a browser. I need to show you something from Partikkel Tek Labs."

Thrown off balance by Jane's apparent lack of intimidation and his own curiosity, Bolton covered his hand then entered a two-digit code into his computer. The office security system display came into view.

"If you open a browser, I can show you proof that the governor knew all along that *the Scrubber* was fake. I talked to the scientist at Partikkel Tek who tested the original design. I have documentation."

Bolton closed out the security system screen. His personal email account appeared. At the top of a list of 20 or so unread messages was one marked URGENT from Anja Hart.

Bolton reacted, "Anja! She's never once emailed me." He opened the message, *Call me after you read this attachment,* was all it said.

Jane looked at Bolton, "You know Anja Hart? She's Yeung's biggest critic. Are you in collusion with her?"

"Collusion? With that bitch? No! She's my granddaughter. Her dad, my asshole son-in-law discovered the protocol that *the Scrubber* is based on. He's working for Bates-Hardy." Bolton clicked on the attachment and watched the progress bar as it downloaded to his computer.

"Your son-in-law is Dylan Baker? He's the guy who stumbled onto the bacteria that eats carbon dioxide and poops out alcohol. Why didn't you tell me? We could have used that information to discredit the governor."

"Don't even bother with Dylan. He's really unimportant. Just a hick who lives in the woods near Seward and guides hunters. He hates me."

"But he has influence with the governor? He also must know the truth about the Scrubber." *A man who is hated by Bolton and trusted by the governor? I need to talk to him,* Jane thought. *I could use Dylan Baker to get to the governor without going through Cheek.*

It was at that moment that Jane knew she could never work for Yeung or anyone like him. No longer could she tell herself that working to put people like Yeung and Bolton into power was a necessary step in her professional development.

A buzzing sound came from the outer office. Bolton brought up the office security cameras on his laptop. On one of the six screens, Jane and Bolton watched Ayushmann Krishnamurthy standing outside the outer office door ring the bell again.

"I need to talk to this guy. You stay here." Bolton closed his laptop and left the room. Jane heard a chair back being jammed against the office door, essentially locking her in Bolton's office.

Jane opened Bolton's laptop to see the login screen. She knew he used a two-digit unlock code, so she entered his lucky number, 22. It was rejected. Jane figured there were 100 possible two-digit number combinations.

She had to work fast. She entered 00, 01, 02, and so on. It wasn't until she got to 24 that the computer unlocked. Looking down at Bolton's numerology card that he had taped to his desk, she saw that 24 was *self-indulgence and arrogance in love.* She instantly felt icky touching Bolton's keyboard. *He uses this to watch porn!*

She selected the security screen that showed Bolton talking to Krish. She turned on the audio for that screen. Krish's high-pitched voice emerged from Bolton's laptop, "You need to do something. She can't be allowed to ruin everything."

"You kill her. You Muslims don't have any morals against killing people." Bolton's voice sounded coldly reasonable.

"I'm no more Muslim than you. You fit the killer profile better than me. When did you ever hear of a Hindu killing children in a schoolyard? No. It's you fucking white Christians."

Bolton paused while digesting Krish's words, "Well. Do you know someone who will kill her?"

"Yes. You. Just ask for forgiveness before you die, and you'll go to heaven—if I understand your religion," Krish spoke boldly.

Jane's hands trembled as she started uploading the recording and the live feed of their conversation to the Super Pac's website. She entered a code to include a live feed from Bolton's office. If Bolton would somehow approve all uploads, anyone viewing the website could see what was happening in Bolton's office.

Jane's next task would be to figure out how to get into the Super Pac's website from Bolton's computer. Once in the website, she might be able to approve all comments and uploads without trusting Bolton would do it. This would get the evidence out in public.

They couldn't kill her if their plotting was public. Jane watched the men walk to the kitchen area of the office. There were no cameras or microphones in there. She would be unable to see or hear them. She wondered if the public would ever see any of the files she had uploaded. Nothing went up on the website without Bolton's approval.

Jane needed to get help. Bolton had her phone, but she noticed a call box for the hotel concierge. She could order room service, get tickets to an event or have the hotel call the police. Jane pressed the call button.

Jane heard the door rattle as Bolton removed the chair from outside. She switched to the login in screen so the upload would proceed in the background.

The men had finished their conversation and were coming for her.

116

RUNNING THROUGH THE WOODS, Reno turned on his high power Cree T6 waterproof headlamp. He cranked it up to full power and the forest lit up like daytime in front of him. The slow-moving motorcycle had a good head start, but its top speed seemed to be a fast walk as it bounced through the woods. Reno could run for hours at nearly twice that speed.

Right now he was in full combat gear and holding a Randall 1911-style pistol built for a left-handed shooter like himself. Tango had told him to pick out any .45 he wanted, so he spent over $8,000 on this one. It had an accuracy of 1.5 inches at 25 yards, but that wasn't based on firing while running in a rainstorm. Reno decided to close within 20 yards, stand and fire. He'd pump seven rounds of hollow points into the back of the fat asshole on the weird bike.

The motorcyclist turned the bike up hill, and it didn't slow down. It might have even increased speed. A moment of doubt entered Reno's mind. He would need to close in fast on his target, or he might get winded. The forest floor was a combination of mud and surface water. This slowed him down as well as an extra 10 pounds from all the water his gear had absorbed during his time in this forest.

At the top of a mild ridge, the bike paused. This was great. Reno decided that after the kill, he'd take the bike back to the others instead of walking. As he got closer he noticed that the bike had stopped at the shore of a recently formed shallow lake in the path. Rocks prevented the biker from going anywhere but into the lake or to wait for death.

As he approached within his desired firing distance, he noticed that the biker had decided to enter the milky-green lake on the motorcycle. That would suit Reno just fine. The bike would stall out, and Reno could get his kill like fish in a barrel.

But the bike did not stall. It seemed to be doing just fine going through 12 to 18 inches of water. The lake appeared to be about 200 yards wide with a rocky bank on the other side. Reno had to get the biker quickly. At 50 yards, Reno stopped and steadied his hand. He fired a double tap, but the bullets were wide to the left. *Easy enough to correct.*

SULLY HAD NEVER been so afraid. A scary guy with a super bright light was chasing him. He couldn't remember what Dylan had told him to do to make the bike shift into a higher gear. The bad guy was gaining. Sully knew the soldier would kill him and take his lever action gun and kill Taylor and Dylan.

Suddenly there was a lake in front of Sully. What could he do? The killer just kept coming. When Sully looked back, he thought he saw a gun in the killer's hand. Easing the Rokon into the lake, Sully was surprised that it seemed to do just fine in the shallow water. It was probably going faster than a man could run in water that deep.

Sully saw two big splashes on his left side just as a double boom came from behind him. The bad guy was shooting, and he was pretty close to Sully.

Two more booms and this time the shots were closer. Sully tasted blood in his mouth. He had bitten his lip, and he knew he was about to die. He suddenly felt the bike go down. It had found a deep place in the lake. It went down on its side, but kept on running, and it floated. *What the hell? A floating motorcycle?*

Sully moved the buoyant bike between himself and the shooter. The next three shots were very close. One of them made sparks off the motorcycle's trailer hitch. He continued to move away from the shooter in the icy water hoping his feet would soon touch the bottom.

The shooter seemed to be looking for something. Maybe more bullets. Sully could hear him cuss. Soon the water became more shallow, but Sully didn't want to tip the bike back up and start riding it because he would be a much bigger target.

Sully felt trapped. He was sitting in 20 inches of icy water with the bike between himself and the shooter. He watched the soldier walk toward him holding his pistol out in front.

Sully knew he was going to die. The killer would eventually just walk up to him and put a bullet in his throat, like he did to Ethan. Then Sully noticed the Marlin's barrel pointing to the sky like a small flagpole. He had never shot a real gun before, but he understood the basics, point and pull the trigger. *How hard could it be?*

It seemed to take forever for Sully to figure out how to unstrap the gun. He was pretty sure all the water had to be out of the barrel in order for it work. Then he worried about the gunpowder getting wet. People talked about *keeping your powder dry.*

As the depth of the lake water increased, the soldier had stopped walking forward and was about chest-deep holding his gun above water, the bright light hurting Sully's eyes. Bad Guy looked like he figured out what Sully was doing behind the motorcycle so he took aim with the pistol.

Quickly, Sully pointed the rifle in the direction of the super bright light and pulled the trigger. The muzzle flash brilliantly lit the lake in front of the rifle. The Marlin kicked in his arms and slammed against his shoulder.

The bullet splashed five feet in front of the killer sending a spray twenty feet out on either side of the splash, but the bad guy's bright light also went airborne spinning high overhead. It hung there a moment and fell, landing about 100 yards away.

Damn, that hurt my shoulder, thought Sully. *Where's the killer? Had he ducked under the water?* Far from where the bad guy had stood, a badly misshapen helmet floated upside down in the lake. Sully was confused, hadn't he missed? Could it be possible that his bullet had ricocheted off the water and hit Bad Guy? *I love surface tension if a ricochet got that killer.*

"That's for you, Ethan!" Sully trembled as adrenaline flooded through him.

Sully waited a few more minutes, but the soldier never came up out of the lake, so he took the motorcycle to the side of the lake to make his phone call. His phone did not survive the trip across the lake. It had flooded, so there would be no call for help.

Shivering, Sully pulled one arm out of his jacket sleeve and put it across his chest. Then he put the empty sleeve over the exhaust of the Rokon. The smelly, smoky exhaust was pumped into his jacket and filled it with hot exhaust. Sully's jacket blew up like a balloon. Water oozed out of it. He kept the sleeve over the exhaust port until his core was warm again.

While he was warming, he decided he should find Taylor and Dylan to tell them no help was coming. For a moment he considered using the motorcycle to find his way to town, but that would leave his friends helpless and surrounded by killers. Sully decided his own safety was less important than his friends.

Once he was warm, Sully continued shivering as he thought about the dangers ahead. His underpants were wet, and his shoulder hurt. He didn't think he could ever use that Marlin again, but he would.

117

"HELLO, THIS IS THE CONCIERGE desk. How may I help you?" the pleasant voice filled the room as Bolton and Krish burst through the door.

"Yes, please send one of your large Alaska salmon pizzas to the penthouse office suite and a growler of Alaska Amber. This is Jane Koss. I'm here with my boss, Colonel Bolton and our friend Krish."

"Of course, Ms. Koss. Anything else?" the voice asked.

"Yes, please send a crew to vacuum the floor of the office suite. It's dusty, especially the closets."

"Certainly, anything else?" the helpful concierge asked.

"No, but please send that cleaning crew as soon as possible," Jane hit the switch to end the conversation. Then Jane turned to face the men. Krish was awkwardly holding an extension cord. He cast a confused look to Bolton.

Jane feigned confidence as she gestured to Bolton's laptop. "Colonel, let's see what Anja Hart sent you. This could be a real break for our campaign and a blow against the governor's."

Looking off balance, Bolton sat down in front of his laptop and brought up his downloads folder. He opened the file Anja had sent him; it was the same report Jane had received from Trevor Calm on the micro-SD drive.

Bolton browsed the document for a few moments, "Christ! This is a 198-page report full of overblown technical language. I'll need to get some of my tech guys to interpret this for me. It could take weeks."

"Here, let me look at it," Jane turned the laptop towards her and quickly found the report's conclusion. She selected text, copied it and then opened a word-processing document. She pasted in the copied text pulled from the report. Then she highlighted parts of it, added some brief editorial comments and returned the laptop to Bolton.

He dutifully studied the page and looked up at Jane, "You are amazing. It's like you were familiar with the document and went right to salient parts."

"When looking at a report like this, you always start with the conclusion. I guessed this report discredited *the Scrubber* and by association, the governor." Jane did not want Bolton to know she'd already studied the PTL report since Calm gave it to her.

"May I see it?" Krish craned his neck to look at the computer.

"No. We don't need you for this," Jane snapped. "I'm running this campaign."

She looked at Bolton, "What's he doing here? Get rid of him."

The Colonel cast a dismissive look at Krish.

Jane went on. "What I can't figure out is why she'd send this to you. This helps Yeung's campaign."

"I have no idea. Maybe this is a fake report," Bolton said.

"No. I'm sure it's real. I spoke with the people at PTL and they said much the same thing," Jane said.

"Don't trust Anja Hart, Colonel. And I wouldn't trust Koss," Krish shook the extension cord in her direction.

"Fuck you, you slimy little weasel. You've never given me good advice," Bolton pointed at the door. "Time to leave."

"You are making a big mistake. Koss will tell you to put that document up on Yeung's website and release it to the media. It's a trap I bet. You'll destroy all our plans. You'll piss off important people and risk jail," Krish spoke as he backed away from Bolton.

"He's right about one thing. You need to put this up on the website instantly, before the media gets it from another source. It's big news. You want everyone coming to our site to get the information. This will bring votes and contributions. Do it now!" Jane shouted like a drill sergeant.

"But what if it's fake?" called out Krish from the doorway.

"It's not. I've got corroborating information in my car. Colonel, you post this and approve any pending posts on Yeung's site. That would be the fastest way to get the files up before the media finds it elsewhere. I'll go down to my car and fetch the evidence I got from PTL. We can eat pizza and plot our next moves."

Jane rushed past Krish, who was on the floor restoring the extension cord and calling out warnings to Bolton. As she left, she heard Krish telling Bolton he was being manipulated by a girl. She guessed Bolton would probably not post the file and upload everything onto the website without seeing her corroborative information, but she could not bring herself to return to the office. She had failed.

In front of the elevator, she walked by an approaching cleaning crew. As she waited for the elevator, tears flowed down her cheeks and her knees felt weak. *They were going to kill me!* Her hands shook as the adrenaline left her body. She hoped she didn't faint before the elevator arrived. Jane knew she had to talk to Dylan Baker.

118

DYLAN STOOD BEFORE ANJA, water still swirling around their feet and the forest full of storm-caused destruction. An uncanny stillness settled over the dripping woods, as if the forest was waiting for something.

Here was the daughter he dreamed he would love and cherish as much as she would love him. She was beautiful and intelligent, like her mother, but she was obviously crazy. Maybe he could help her. Maybe somehow he could help her see reality.

"Anja, don't you see your expectations are completely unrealistic?" Dylan said, a catch in his voice. As he spoke, several porcupines waddled past the group, completely ignoring the humans. Part of Dylan's mind tried to figure out why these shy animals were acting so strangely. *Could it be the glacial melt water? The glaciers are melting?* The rain had stopped, but milky-green water ran down the hillside.

"Where did you send the text?" Anja said ignoring Dylan's statement.

Roger pointed his Colt M4 assault rifle at a spiky animal and fired a 15-round burst. The 5.56mm fully jacketed rounds spat out at 16 rounds a second. In less than a second, the shredded porcupine was blasted across the clearing.

"Knock it off," said Phoenix "You are firing your weapon outside the parameters of the mission."

"Fuck you. It's my gun and you're not my CO. Plus that's an invasive species, just like humans," Roger said. He looked to Anja for approval, and then gazed around seeing dozens of animals running and walking downhill toward Resurrection Bay.

Phoenix searched for Tango, but she had left the brightly lit area in front of the high rocks and disappeared into the darkness.

"Neither porcupines nor humans are invasive in Alaska," said Dylan.

"Fuck you, too," said Roger whose formerly contained hostility seemed to be unleashed out in the woods.

"Dylan, I promised you and your scientists safety if you would tell me to whom you texted that report," Anja said impatiently.

A feminine cry came from the darkness. All the armed men turned toward the sound. They could see Taylor walking into the clearing, her hands up, as Tango walked behind her.

"Bolton," said Dylan. It was just a guess, but he wanted to get Anja's attention. "I texted it to your grandfather. Are you going to kill him, too? Are you going to wipe out what's left of your family for an impossible plan?" said Dylan. "It makes no sense."

Anja ignored him, pulled out her phone and cursed. No bars. She entered a text to someone at her office, *Get Bolton.* It would send as soon as her phone got a signal. She looked up at Dylan.

"The earth is my family. I meant it when I said I didn't want to kill anyone. I really don't want to kill you. Although you mean nothing to me emotionally, I know you were the sperm donor in my life. I understand that you want something from me that I can't give."

"But you can. Let's take some steps to a healthy father/daughter relationship," urged Dylan.

"Now who's unrealistic? Can't you see? I would like you to understand that you and your scientists need to die in order to help us save the earth. It's a completely noble death for a noble cause," Anja explained in a calm, reasonable voice as if she were telling him to order the meal combo because it came with fries.

Anja signaled to Tango. It was the signal that meant Tango was to kill Dylan and Taylor. Anja possessed unshakable faith in her beliefs, but she did not want to watch the execution. She paid Tango to do that.

Walking away from the High Rocks area to the forest, she called to Roger and Andrew to come with her. Roger looked disappointed. He wanted to watch.

Phoenix and Tango roughly pushed their victims against the steep, rock wall then backed up and raised their weapons.

THE MILLENNIAL storm had passed leaving the super -saturated earth around the Kenai Fiords a chance to rest. Although the rain had stopped, it was too late to save the Harding Icefield's lake. The super-cooled water flowing through cracks in the ice dam changed from a trickle to a river. Friction caused by liquid water flowing under the dam did what water always does to ice, it makes the ice float and rise up allowing the lake to flow downwards towards Resurrection Bay.

A mile away and two thousand feet higher in elevation from High Rocks, a cataclysmic icy explosion of liquid water burst from the side of the icefield. Super-cooled air, denser than the warm air around it, began to sink. This created a powerful katabatic wind gust blasting downward ahead of the water.

The outflow of millions of gallons of water shook the rocky spine of the icefield down to the core of the formation.

Sympathetic vibrations activated sleeping tectonic plates. Pressure forced the Continental plate upwards over the Oceanic plate. This had last happened in 1964 and caused the shores of Resurrection Bay to drop nearly six feet in elevation. This time it produced a slow and steady rumbling that spread out over south-central Alaska.

119

A MILE FROM HIGH ROCKS, and following the forest-flattening katabatic wind gust, a green wall of water burst from the icefield. The flood easily pushed house-sized boulders ahead of it and tumbled down the mountainside toward Resurrection Bay.

SULLY POINTED the Marlin at Tango, but his first shot hit the hard, granodiorite cliff behind Dylan and sent scores of quartz projectiles into Dylan and Taylor causing tiny cuts all over their faces. The second shot was higher and entered the minivan-sized ledge above their heads. Tiny rock shards blasted out of the ledge and peppered Tango and Phoenix, who were already seeking cover by some greywacke cliffs to the side.

The third shot went even higher and hit a tree dropping a baseball-bat sized limb near Tango and Phoenix's cover.

Out of ammo, Sully left his hiding place in the woods and ran to Dylan and Taylor holding the Marlin, "I'm out of bullets! I'm out of bullets! Do you have any more?"

"What's going on?" yelled Roger as the he emerged from the trees flanked by Anja and Andrew. His voice sounded uncharacteristically panicked. He had sunk up to his knees in the earth as the super-saturated soil became liquefied under the trembling of the mountainside.

Trees rose up out of the earth and toppled, and Taylor and Sully screamed when they saw Andrew disappear under the earth as if he were swallowed by quicksand.

Dylan pointed to the ledge above his head, "Up!" He screamed. Pushing Taylor and Sully up the rough granodiorite walls. The skin-shredding rough wall allowed them traction as they moved upward.

Near the trees and waist-deep in the earth, Anja yelled, "Help me! I'm sinking!"

As Dylan struggled to get Sully's bulk up the cliff face and onto the ledge, he yelled to his daughter, "Anja, I'll save you."

Sully slipped, and Dylan again had to get on his knees and push with his back to help Sully up.

The sky above them turned dark, filled with airborne forest debris blown on the massive wind gust ahead of the descending flood.

Once Dylan could see that Taylor was helping Sully up to the ledge, he pulled the rope from his pack, looped one end on a rock then threw the other toward Anja. The wind took the rope away from her grasp.

"Help me!" Anja's panicked voice sounded high and childlike in the animated forest.

The earth shook. Trees collapsed. A bear skull rose out of the earth, along with anything less dense than the mud. "Daddy!" screamed Anja as she sunk below the ground.

"Dylan! Get up here!" yelled Taylor.

He glanced over to where Tango and Phoenix were attempting to climb the greywacke rocks and dodge the falling stones all around them. That was when a roaring green wall of water burst against the High Rocks, sending a shudder through the hard 100-foot wide seam of intrusive rock. The surrounding soft shale and greywacke disappeared under the powerful onslaught.

From the ledge and above the violent lake water, Sully, Taylor and Dylan marveled at the angry, frigid waters as they scoured away the forest and soil. Being on the lee side of the aquatic wave, Dylan had time to rope up his team to prevent the vacuum created by the floodwaters from sucking them off the rocks and into the maelstrom.

HOURS LATER, Dylan trembled as he tried to come to terms with what just happened. The entire forest on his mountain was gone. Much of the mountainside was also gone, washed into the bay leaving an unrecognizable moonscape behind.

He lost his home, his daughter and his mountain. A deep, stunned sadness swept over him as tears ran down his cheeks. He watched a huge brown stain spreading over the bay as the liquefied mountainside became part of Resurrection Bay.

Dylan turned to untie himself from his companions. Sully broke the spell when he held out a soggy granola bar and asked if anyone wanted a bite.

120

JANE CLIMBED INTO her X90, set her GPS to Seward and pulled out of the Netsvetov Hotel parking garage.

Upstairs, Bolton paused in his argument with Krish to look at his phone. "I just got a text from Dylan. It has an attachment."

"Don't open it. It's probably a virus and it will destroy your phone or worse." Krish leaned over Bolton to see his phone.

"This was texted to me and to his wife. She's a journalist. He wouldn't send her a virus." Bolton opened the attachment and started reading.

"What is it? What does it say?" asked Krish, who seemed disappointed that Bolton's phone didn't show signs of hacking.

"Jesus! This is a document written by Dr. Sullivan Meyers confirming everything Jane said about Anja's document."

"It's a trick. Why would Anja and Dylan want to let out information that would get Yeung elected?" Krish said.

Bolton ignored Krish and blurted out, "I bet Jane is right. As soon as Dylan's wife reads this, she's going to write a story about it."

"If she releases the story before you, no one will visit Yeung's website to see the original citation." Krish started to get excited.

"Damn, you are right. I got to get the document Jane wrote and this text from Dylan up on the website right away. Then I need to hold a press conference getting all the Alaska media to go to the website." Bolton opened his laptop grinning as he imagined the destruction to the governor's campaign he would cause.

As Bolton uploaded Jane's announcement, he noticed all the video files awaiting his approval. "What's all this?" Bolton sounded suspicious.

"Those are from Jane's computer," Krish pointed to the files. "They are probably the proof she said she had."

Just then the door out side buzzed. "That's probably Jane with more proof," said Krish. "I'll let her in."

That was all Bolton needed, he clicked the box, *Approve all comments.* Jane's video files and attachments were sent out to the world just as Krish walked in with pizza and beer.

HOURS LATER, Jane was thankful her small 4-wheel drive X90 could skirt all the storm debris on the highway into Seward. She heard on the radio about a huge landslide out beyond Lowell Point, but Seward itself only suffered some washed out bridges and minor flooding.

IN THE SEWARD Senior Center, Taylor put Dylan's phone on his cot. He opened his eyes and looked confused. "It's charged now, actually works, and you have a new message," said Taylor.

Dylan looked at his phone with dread. It seemed to only bring him bad news. He could barely allow himself to open his email. It had been a harrowing journey from their perch on the granodiorite ridge to the town of Seward, and he did not need any more loss in his life.

Fortunately, the flood had mostly done its damage in the Tonsina Creek Canyon and caused little damage in Lowell Point and nearly none in Seward. The local authorities were gathering the few refugees in the Senior Center until better arrangements could be made.

Sully walked in with a guest, who appeared to be an Asian teen of indeterminate gender, and a handful of huge cookies, "Hey, they just put out cookies and soup. I got the last chocolate chip cookies, but there are still some oatmeal left in there."

Dylan smiled, looked at his phone then appeared puzzled. The urgent text message was from Bolton. He wanted Dylan to check the Super Pac's website.

"Hey Dylan, know who this is?" Sully elbowed the slender visitor.

"No. I've never met this person," Dylan didn't want to assign a gender to the visitor until he had more information.

"This is Jane Koss, Yeung's Super Pac administrator.

Dylan stood, "What are you doing here?"

"Mr. Baker. I've resigned from the campaign. I came here to talk to you." Jane stepped forward and suddenly did not look anything like a teen.

"Me? How do I know you are not here at Bolton's bidding?" Dylan unconsciously rubbed a bruise on his wrist. "Bolton just messaged me and asked me to call him. It's pretty interesting that you just happen to drop by when he asks to talk."

"I understand your concern. It's justified given all the dirty tricks Bolton has pulled. But he's also pulled some fast ones on me. Let's talk before you call him. I can suggest some questions for you to ask."

Jane sat down on Dylan's cot and the two started a deep conversation while Sully and Taylor wandered into the office to see if they could borrow a laptop.

A few minutes later, Sully came back and made a sign for Dylan and Jane to follow him. Soon they were in the senior center office looking at the monitor of an older Dell desktop computer. On the monitor was a grainy picture of Bolton in his office. He appeared to be pouring himself a drink.

Dylan looked at the picture. "You're shitting me. This is a live picture?"

"Yes, he doesn't realize he approved a live video/audio feed from his office. Call him," said Jane.

Fascinated, Dylan punched in Bolton's number. A few moments after he could hear Bolton's phone ring through the earpiece, he saw Bolton pull his phone out of his pocket.

Bolton stared at his phone and waited until the fifth ring before answering by pushing his speakerphone button. Bolton pretended he didn't know it was Dylan on the other end of the call.

"Yes, who is this?" Bolton said holding the phone out in front of him.

"It's Dylan Baker. You wanted me to call you."

"Yes. I got your text. It was very interesting. I wanted to know your price for the original documents." Bolton sat back in his big chair.

"My price? The original is gone. Burned up in a house fire," said Dylan.

"Too bad. I have a big press conference in ten minutes and I wanted to claim I had the original report. I already released two different reports based on the PTL documents, one from a reputable scientist named Dr. Sullivan Meyers and another from Yeung's Super Pac administrator. Once the Alaska news media analyzes this report, the governor will be unelectable. The public will view her as someone who doctors up evidence to trick voters."

"Listen Bolton, you know the governor had nothing to do with doctoring up evidence. She was a victim of a very clever scheme run by a sophisticated environmental terrorist organization headed by your granddaughter." Dylan choked for a moment, *And she was my daughter.*

"I don't care about that. It doesn't matter that the governor is truly innocent. What matters is that voters think she's guilty. We can make her look like she knew about *the Scrubber* being a fake all along."

"You would lie to voters?" Dylan asked.

"That's what Super PACs do, lie," laughed Bolton. "We're going to make the Mountain Club look bad, too."

"But they didn't know what Anja was up to. She was running a terrorist organization while pretending to work for Mountain Club. They are guilty of trusting their executives, but they did nothing wrong."

"Once again, the truth doesn't matter. What matters is that we get Greg Yeung elected. My friends and I bought some drilling rights that we want to exercise once Yeung deregulates public lands."

"Would your friends include Nigerian and Saudi and other oil mafia types?" Dylan asked.

"How did you find out about that? I needed their money to combine with my own. If you publicize my relationship with them, I'll just deny it." Bolton looked like he had lost some of his bluster.

Jane held up a notebook for Dylan to see. He read it and smiled.

"What about manipulating the stock market to get money for this investment?" Dylan smiled.

"Now just a god damned minute. You could never prove that. I've hidden my tracks too well. I know how to manipulate the system."

"And what about taking money from Yeung's Super PAC to cover your losses?" Dylan tried not to smile.

"None of my loans from the Super PAC can be traced. The FBI are fools. How did you find out about this stuff? None of my people would dare to leak anything!" Bolton shouted.

"Never mind how," said Dylan. "Why did Dave White go in on the con. He was an important energy researcher."

"My guy, Krish ferreted out the answer to that. Dave White was promised he'd get a huge research grant from Mountain Club. He really believes he can build *the Scrubber* if he just had some funding. In his own mind, he was just stretching the truth, and not lying. Now you tell me, how did you find out about my stock activities. Only my partners knew about that."

Dylan handed the phone to Jane. "Hi Colonel."

"Jane? What the hell are you doing in Seward? You get back here right now. Quit telling tales to that idiot mountain man."

"What I am doing here? I'm streaming this call live on Greg Yeung's YouTube station. We already have over 300 shares. Soon, everyone in Alaska will watch this," Jane said.

"Bullshit. I control that site," said Bolton.

"Actually, you uploaded a live feed of your office cams to the website when you posted my paper," Jane said. "Twenty minutes ago, I informed the FBI about you and your activities as well as all the self-incriminating material you put on our site. Bolton, you are financially ruined, and you will go to jail."

On the computer monitor Dylan watched two dark-suited men rush towards Bolton from behind. One was holding handcuffs. Bolton turned to face them. "Hey! How the hell did you get in here? This is a private residence."

"Colonel Bolton? Colonel William Bolton? FBI. You are under arrest."

121

THE MUFFLED CITY sounds of Anchorage competed with coffee house clamor as Jane Koss took the seat at the table reserved for Governor Leola Bates-Hardy. An aide indicated that the governor would return momentarily.

Jane took a deep, relaxing breath and realized how much she appreciated a few moments to collect her thoughts. A lot had happened during the last couple of days. Her falling out with Bolton had left her with an opportunity to decide what was important to her. Helping Dylan snare the egocentric Bolton felt like righting a wrong. Now Jane found herself at another fork in her life path.

Governor Bates-Hardy, exuding a purposeful yet friendly air, slid into the seat next to Jane. "Thank you for seeing me, Jane. Have you had a chance to review my proposal?"

"I have, Governor. Thank you for considering me." A position on the governor's staff would have many of the ideal characteristics Jane had searched for in the past. Creating political strategies from the camp of the superior politician was a wonderful job on the road to her dream of running a presidential campaign.

Before Jane could vocalize her thoughts, Hardy continued, "I would also like to thank you for bringing the real scrubber performance data and Bolton's behind-the-scenes activities into the public arena."

THE GOVERNOR's authenticity and integrity again impressed Jane. The creative energies around the governor were not unlike those positive traits that surrounded Bolton—the ability to see what is possible with only the building blocks at hand was a trait Jane admired.

"I've decided to take a break from politics, Governor. I surprise myself as I say these words. Had you asked me a year ago about my dream job, a position on your staff would have been near the top. But, lately I've disappointed myself. I was too quick to rationalize and justify the things happening around me. I need a little introspective time to understand how I got so far off my course."

"Still, your strategic skills would have been a wonderful resource for my team."

"Governor, you know that I'm a bit of a liability at this moment."

"But that would pass in a relatively short period of time. Leaders have endured much more and gone on to make significant contributions."

Jane knew the governor was right. Within hours of being unemployed, the staff at the Netsvetov Hotel had offered Jane a director position on their strategic planning team. Then the Anchorage Gazette offered to publish a more detailed series on the behind-the-scenes activities of Bolton and company. *No thank you!*

The news announcer on the coffeehouse flat-screen TV was summarizing the latest updates on "Bolton-gate". Jane returned her attention to the governor. "Aren't you scheduled for a teleconference later this afternoon? What are you going to tell the public?"

Hardy seemed surprisingly calm. "I'm going to continue to emphasize my record of public service. I'll give them the truth. I always do."

Governor Bates-Hardy's gaze focused intensely on Jane. "I understand there are several state-level, anti-carbon Super PACs in various stages of getting their acts together. If you are interested in working for any of these organizations, I'd be happy to provide a recommendation."

In that moment, Jane realized that Governor Bates-Hardy was the type of superior leader she aspired to work for.

Epilogue

SULLY, CONRAD, TAYLOR AND DYLAN waited outside the doors with a crowd eager to get into the Miners and Trappers Ball. Every March the entire state of Alaska seemed to gather in Anchorage to celebrate the Anchorage Fur Rendezvous and the ball seemed like the right activity for friends, with live music, a beard contest and plenty of food.

"I think I'll win the beard contest," announced Conrad. "I haven't shaved since I left Seattle."

"But that was yesterday," said Sully. "Anyway, we didn't come up here to win a contest, we came up here to celebrate with Dylan."

Dylan looked pleased while dressed up in his new flannel shirt, suspenders and hiking boots. "Thanks, guys."

"We also came to present a bid on an ice pack study that newly re-elected Governor Bates-Hardy announced two months ago. Conrad's consulting work with Microsoft will be over soon, so he can lead the project. So it worked out that we could see you. Where's your wife? Is she still in Africa?" asked Sully.

Dylan's face revealed a pained expression.

Taylor interrupted the talk, "Dylan and I are going to get some coffee. Can you guys hold our places?"

"Sure, bring us some donuts," said Conrad.

Stepping away from the line, Taylor and Dylan crossed over to a snack cart and placed an order.

"So Dylan. Tell me how are you doing?" The intensity of her look signaled her wish for a complete, honest answer.

"When I lost everything, my Seward friends came to the rescue. Then my part-time summer job with Fish and Game doing field surveys of wildlife populations expanded and they offered me a position based here in

Anchorage. The other good news is that I'm thinking of opening a guiding business in the Anchorage area."

"Nice outcome. So why did you flinch when Conrad asked you about Suzie?"

Taylor could see him exhale and a trace of sadness crossing his face. "She's an amazing person, and she will do what she wants. But she has become a different person in a world that I can't even relate to. She's going to live in Africa with some English doctor. But I am attempting to move on now, both physically and emotionally."

Conveying her empathy, Taylor touched Dylan's arm for a moment. "Please call me if you need to talk to someone." Dylan acknowledged her offer. They collected their order and started back to the others.

"What are you up to these days?" asked Dylan politely.

"I got a research position at University of Florida in the microbiology department. It's temporary, but I hope it will become permanent. I love being a Gator."

"I'm really happy for you, Taylor. It seems like you are on the right path to reach your dreams," Dylan grinned. "Whatever happened to Dave White? Is he in jail? I heard that Bolton will be in prison a long time."

As they stepped back in line Sully complained, "You only got four donuts? What about everyone else?" Sully grabbed a chocolate glazed. "Hey, this is the Miners and Trappers Ball, maybe we'll meet some miners."

"You're too old for minors," joked Conrad.

"No. I mean maybe that guy who makes the great pancakes, John Jolly and his guys are here."

"I doubt if they are here in Alaska in March. They're still enjoying warm Hawaii now," said Dylan as they approached the front entrance.

"Dave White and his dream of burning fossil fuels with no environmental cost are also gone. Both White and the dream have dropped off the map and into a minimum-security prison. It was a great dream. I bought into it, but it looks like it will never come about; at least until some of the problems can be worked out," said Sully.

"Don't be too quick to discount *Ralstonia* and researcher creativity," said Conrad. "Labs in Europe and the US have new strains of bacteria that are more efficient at converting CO_2 to biofuels."

They could hear the band strike up a country western dance song from inside the hall. "So there are some great women here?" asked Conrad. "I'm an expert at dancing on one foot." He twirled around on his good leg.

"You might find someone as cute as that Jane Koss in there. She's working for the *Save Alaska* Super Pac. It supports pro-environment political candidates. Maybe she can make headway on getting us Alaskans to realize we need to lower our carbon footprint," said Dylan.

"She's too tall for me," said Conrad. "I was thinking of someone like Taylor, but seated in a very low chair." Conrad gave a friendly wink to Taylor.

"Did you guys know that I got a temporary job at University of Oregon?" said Sully. "This job opened when they moved someone up to take Dr. Carlyle's position. My office is near the student union, great food! You should try their spicy tater tots."

As the group passed their tickets to a pretty girl dressed in a skimpy lumberjack costume and entered the room full of locals and friends ready to party, Dylan decided he would enjoy the moment.

About the Author

- Black belt in Shotokan Karate
- Honored by the U.S. National Teachers' Hall of Fame
- Spent two summers as a kayak guide in Seward, Alaska
- Has traveled the world writing and photographing for motorcycle magazines
- Has published scores of articles in mainstream print magazines and professional journals, under a different name
- Has a total of 5 graduate and postgraduate degrees in subjects ranging from microbiology to history
- Plays on an elite Ultimate Frisbee team
- Is author of three other books under different names
- Speaks fluent Spanish and German
- Has traveled throughout North and South America, Asia and Europe
- Currently has three screenplays available from Mt. Hood Press

Tyler Blackthorne, as you've figured out, is a pseudonym. The Portland, Oregon father-daughter writing team of Bruce and Kelly Hansen was joined by Minnesota scientist/writer Bob Young to write the first Dylan Baker book, *Denali*. *Arctic Forces* is a creation of Bob Young and Bruce Hansen. Readers can find out more about this writing team by visiting our website, *mthoodpress.com*

About Tyler Blackthorne's Writing

If you enjoyed this tale, please take a moment and post a review on Amazon.com or your favorite book review website.

Bruce Hansen, Bob Young and Kelly Hansen want to thank our family and friends for all the support to make this book happen.

You might also enjoy, **Denali**, by Tyler Blackthorn. It tells the tale of Tango's first meeting with Dylan in the Alaska wilderness.

The third Dylan Baker tale, **Arctic Protocol**, tells the tale of what happens when Dylan decides to lead a wilderness adventure trip on an Alaskan mountain full of killers desperate to keep the location secret.

Ask about our screenplay version of Denali and other tales by contacting: info@mthoodpress.com.